Writing a Professional Life

THE ALLYN AND BACON SERIES IN TECHNICAL COMMUNICATION

Series Editor: Sam Dragga, Texas Tech University

Thomas T. Barker

Writing Software Documentation: A Task-Oriented Approach

Deborah S. Bosley

Global Contexts: Case Studies in International Technical Communication

Paul Dombrowski

Ethics in Technical Communication

Laura J. Gurak

Oral Presentations for Technical Communication

Dan Jones

Technical Writing Style

Charles Kostelnick and David D. Roberts

Designing Visual Language: Strategies for Professional Communicators

Carolyn Rude

Technical Editing, Second Edition

Gerald J. Savage and Dale L. Sullivan

Writing a Professional Life: Stories of Technical Communicators On and Off the Job

Writing a Professional Life

Stories of Technical Communicators On and Off the Job

Edited by

Gerald J. Savage
Illinois State University

and

Dale L. Sullivan
Michigan Technological University

Allyn and Bacon

Boston London Toronto Sydney Tokyo Singapore

Vice President: Eben W. Ludlow
Editorial Assistant: Grace Trudo
Executive Marketing Manager: Lisa Kimball
Production Editor: Christopher H. Rawlings
Editorial-Production Service: Omegatype Typography, Inc.
Composition and Prepress Buyer: Linda Cox
Manufacturing Buyer: Suzanne Lareau
Cover Administrator: Linda Knowles
Electronic Composition: Omegatype Typography, Inc.

A Pearson Education Company
160 Gould Street
Needham Heights, MA 02494

Internet: www.abacon.com

Between the time website information is gathered and then published, it is not unusual for some sites to have closed. Also, the transcription of URLs can result in unintended typographical errors. The publisher would appreciate notification where these occur so that they may be corrected in subsequent editions. Thank you.

Library of Congress Cataloging-in-Publication Data

Writing a professional life : stories of technical communicators on and off the job / Gerald J. Savage, Dale L. Sullivan.
p. cm. — (The Allyn and Bacon series in technical communication)
Includes index.
ISBN 0-205-32106-2 (pbk.)
1. Technical writing. I. Savage, Gerald J. II. Sullivan, Dale L. III. Series.

T11 .W74 2001
808'.0666—dc21

00-055844

Printed in the United States of America

10 9 8 7 6 5 4 3 2 1 05 04 03 02 01 00

To Sue and Sheryl

Contents

Topical Contents

Authority, Ethos, and Identity

Balancing Personal and Professional Life

Collaboration and Teamwork

Consulting and Contracting

Definition of Technical Writing

Draft Reviews

Education

Ethics

Gender Issues

GUI Design

Internships

Interpersonal Skills

Jobs, Finding or Changing

Office Politics

Professional Standards

Professional Status

Project Management

Relationship between Professional Cultures

Rhetorical Strategies and Style

Software Documentation

Technical Communication and Training

Transition from Academia to Professional Practice

Varieties of Technical Communication

Writer–Editor Relationships

Foreword by the Series Editor

The Allyn & Bacon Series in Technical Communication is designed for the growing number of students enrolled in undergraduate and graduate programs in technical communication. Such programs offer a wide variety of courses beyond the introductory technical writing course—advanced courses for which fully satisfactory and appropriately focused textbooks have often been impossible to locate. This series will also serve the continuing education needs of professional technical communicators, both those who desire to upgrade or update their own communication abilities as well as those who train or supervise writers, editors, and artists within their organization.

The chief characteristic of the books in this series is their consistent effort to integrate theory and practice. The books offer both research-based and experience-based instruction, not only describing what to do and how to do it but also explaining why. The instructors who teach advanced courses and the students who enroll in these courses are looking for more than rigid rules and ad hoc guidelines. They want books that demonstrate theoretical sophistication and a solid foundation in the research of the field as well as pragmatic advice and perceptive applications. Instructors and students will also find these books filled with activities and assignments adaptable to the classroom and to the self-guided learning processes of professional technical communication.

To operate effectively in the field of technical communication, today's students require extensive training in the creation, analysis, and design of information for both domestic and international audiences, for both paper and electronic environments. The books in the Allyn & Bacon Series address those subjects that are most frequently taught at the undergraduate and graduate levels as a direct response to both the educational needs of students and the practical demands of business and industry. Additional books will be developed for the series in order to satisfy or anticipate changes in writing technologies, academic curricula, and the profession of technical communication.

Sam Dragga
Texas Tech University

Preface

The lack of firsthand stories from technical communicators themselves, and the growing recognition by our profession's scholars of the importance of such stories to students and to the professional culture, suggests the need for narratives specific to this culture. We believe that this collection of stories, the first of its kind in our field, begins to fill that need. The issues, struggles, values, and meanings represented in each of the twenty-three stories in this collection both reinforce and challenge a number of the current "truths" of the profession.

Reflecting the state of the technical communication field at present, many of these narratives are situated in computer-related industries, and most of the authors are women. We believe it is worthwhile to include multiple stories from the same general sector of practice because they afford us the opportunity to see how standard practices may be manifested in various ways. We see in these stories, as well, a number of significant differences in corporate cultures and in the writers' perspectives on their work. But the collection includes stories from writers in a wide range of practices—medical writing, freight industry safety, editing, marketing, civil engineering, and R & D.

The stories we have collected fall into three broad categories: stories about getting started in the profession, stories about the technical communication process, and stories about the lives of technical communicators beyond the job itself. Each of these categories is characterized by certain themes and issues. Initiation stories, for example, commonly reveal concerns about the transition from academia to professional workplaces. Several of these stories portray internships; others capture the challenges and triumphs of job hunting and the discovery that much remains to be learned on the job that was not taught in school.

Stories of the technical communication process dramatically portray a number of abilities and principles that are typically covered in writing courses: audience concerns, collaboration and teamwork, project management, the pressure of deadlines, and the importance of interpersonal skills. But they also call attention to some realities that are largely overlooked in the classroom: the increasing importance and unique challenges of working as a consultant or contractor, the challenges and tensions of working effectively with subject matter experts (SMEs), the complexities of the draft review process, and the difficulties and rewards of maintaining professional standards.

Stories of life beyond the job reveal the ways in which personal life and professional life can become intertwined. Issues of identity emerge in such narratives, and, of course, the problem of balancing personal and professional commitments and interests. Interestingly, it is particularly in these stories that we see the writers trying to define technical writing.

Certain themes and issues pervade all three categories of stories, although they are addressed from a variety of perspectives. Relationships between technical writers and subject matter experts such as programmers and engineers appear in a number of stories in every category. In some cases the relationships are cooperative; in others they are difficult, even hostile. In most cases, whether the relationships are warm or frigid, writers and SMEs seem to be working in a cultural border zone, reminding us of the "two cultures" issue raised by C. P. Snow forty years ago. This issue is seldom addressed in our current research or textbooks, but it clearly affects the daily lives of technical communicators.

Several stories in each category show a preoccupation with interpersonal skills and the challenges of working closely with other people, even when such work may not involve formal teamwork. Aspiring writers who have imagined technical communication to be a cloistered practice will be surprised to learn how much of the work involves building good relationships with other people.

A number of issues concerning the professional status and recognition of technical communicators emerge in these stories. Beginners are concerned with establishing their own status as professionals, and veterans tell about hard-won battles for professional standards and recognition, fought not simply for their individual careers but for the profession as a whole.

Office politics is another common theme. Although we often consider it a trivial matter, the politics of the workplace colors many circumstances and issues of professional life and takes on more than trivial significance. Clearly, these pervasive themes and issues cannot be entirely separated from each other, and it is instructive to see the many ways in which they combine from story to story.

We hope that *Writing a Professional Life* will be read and talked about by veterans of the profession as well as by students and academics. A number of new questions about the nature of technical communication practice can be articulated in response to these narratives—questions that may hold promise for new directions in research. We hope and expect, as well, that they will stimulate further stories to inspire, amuse, startle, and trouble us and will play an essential part in the continuing growth of the technical communication profession.

Acknowledgments

This book is the culmination of many conversations not only between the two editors but also among colleagues and friends who encouraged us to see it through despite the many demands for our time. In particular we would like to thank Teresa Kynell, who was generous in sharing her own publishing experience, in encouraging some of the writers to contribute to the collection, and in reassuring us in a moment of doubt.

Our appreciation of the contributors to the collection goes beyond the work they contributed. We thank all of them for their patience through sometimes prolonged silences on our part and for their enthusiasm for the project.

We would like to thank Sam Dragga, Series Editor of the Allyn and Bacon Series in Technical Communication, for his early interest in the project, his warm enthusiasm for the prospectus, and his advice as the collection developed. Eben Ludlow, Vice President of Allyn and Bacon, gave us considerable encouragement and procedural advice early in the project and has been responsive throughout. Editorial Assistant Grace Trudo has been a constantly reliable resource and facilitator, providing information we needed and helping us to keep the project moving. Kristine Maleri, our Allyn and Bacon representative, was most helpful in facilitating initial contacts and was supportive of the book before it was fully launched as a project. Both the copyeditor and the staff at Omegatype Typography provided gracious, insightful, often inspired copyediting. It has been a pleasure to work with this publishing team.

We would also like to thank the following reviewers: Bernadette Longo, Clemson University; Cezar M. Ornatowski, San Diego State University; and Richard Johnson Sheehan, University of New Mexico.

It is customary to reserve the final thanks for spouses and family. Custom, however, does not allow adequate expression of the appreciation we feel for the love and support we too often thoughtlessly count on and constantly receive from our families.

I would like to thank Sue Savage, who has given encouragement and considerable practical help with the project. I would also like to thank Sarah Ellis, Jennifer Blanchard, Percy and Sue Flotte, and Betty Savage, whose enthusiasm for the book and interest in the process have made pleasant a process that could have been problematic due to divided commitments.

—GJS

I would like to thank my wife, Sheryl, and my children, Blaise, Ember, Phillip, and Becky Sullivan, who supported me during this project and gave up opportunities to do things together so that the work could be done.

—DLS

Writing a Professional Life

Part I

Initiation Stories

Technical communicators enter the profession through various doors. Most of the writers of the following eight narratives were students preparing for careers in technical communication just before they started the internships or jobs their narratives portray, but beyond that common experience, their stories diverge. Some of the writers' work experiences are good, and some border on the nightmarish. These stories reveal work environments and how they contrast with expectations developed in school as well as the often difficult adjustments to long hours, unpredictable changes in projects, and coworkers who don't appreciate or understand technical communication. The stories also show how humor, commitment to professionalism, the ethic of hard work, and the courage to persevere—or to quit—result in rewards and professional satisfaction. These narratives depict writers struggling to enter the profession, struggling to stay in the profession, struggling to transform themselves and their understanding of what technical communication means, and struggling to renew their sense of self-worth as they make difficult career changes.

The first two narratives describe internships, but from quite different points of view. Rosalie Young Dwyer's story, "I'm a Nibbie: The Tale of an Object-Oriented GUI Developer," begins in an internship and concludes in a job she loves, though she wonders in the end whether she's still a technical writer at all. Her narrative raises the question of what writing involves as technologies transform not only the media in which we write but the nature of writing and the function of texts. In "The Great Pyramid War," Brad Connatser gives a manager's perspective of an internship in an R & D company. What should have been an excellent internship turns sour, leaving the manager, and the reader, wondering how much of the blame is his own, the student's, and her teachers'. Michael Asay's "Glimpse into Reality" takes us through a day in his internship with a software company where the schedule is flexible, the SMEs respect his writing skills, and the lunches are free. He describes what would surely be the ideal internship or job for most technical communicators; only the commute is a nightmare.

In "My Entry-Level Life," Kendra Potts maintains a sardonic humor about her work culture, in which, despite its many unique aspects, many of us may think we recognize people we have worked with ourselves, from the sexist manager to the consultants with color-coordinated business cards and hair.

Melissa Alton, Alina Rutten, and Beth Lee all tell stories about the struggle to prove themselves as writers and as valued members of the professional teams in their companies. In "First Time Out," Alton discovers technical communication accidentally, after a disastrously brief career as a high school English teacher. Although she goes back to school to learn this new profession, she argues that technical communication is really a kind of teaching. We might ask in what ways all of the stories in this collection confirm or deny Alton's argument. In contrast to the troubles faced by some other beginners, Alina Rutten, in "How I Became a Goddess," has only to overcome her own self-doubt as she confronts the technical, rhetorical, and deadline challenges of her first job and turns a market loser into a winner.

In "A Job like a Tattoo," Lee is sure she's a technical writer when she begins her first job but comes to question her competence after a series of confrontations with a hostile programmer. This is one of several stories we will read that raise concerns about misunderstandings among professional cultures.

In "It's Not Mark Twain's River Anymore," Carol Hoeniges comes to realize that she's a technical writer who doesn't particularly like high technology. She is forced to face the fact that working in software documentation will never fulfill her professional goals. She moves to a small consulting company in which her work as a writer of transportation industry safety manuals can help save lives. The challenge is not just in understanding the technology but in understanding and being accepted in the workplace cultures in which her documentation must function.

Rosalie Young Dwyer

I'm a Nibbie

The Tale of an Object-Oriented GUI Developer

Rosalie Young Dwyer is a user interface developer in Chicago, Illinois. She received her bachelor of arts degree in English from Seton Hill College in 1992 and her master of arts degree in the teaching of writing from Illinois State University in 1997. She lives in Aurora, Illinois, with her husband, David, and her daughter, Dana.

What in the world is a "nibbie"?

I had just begun my out-of-school career as a technical writer when I found myself presented with the opportunity to become a builder of NIBs: a "nibbie." Let me explain.

Like many English students, I had sat through courses in technical writing, wondering if I would fit in the field. Could I find professional fulfillment in documenting wacky software or writing a manual for operating a hair dryer? After getting my master's degree in composition and rhetoric and plodding through two semesters of a doctoral program, the call to get a real job and make some money grew too strong to ignore; near the end of the spring semester, I e-mailed an old grad school friend who worked for a company in Chicago, asking her if the tech writing department needed any interns for the summer.

Since becoming a nibbie, I've adamantly stuck to my belief that I am no longer a technical writer, per se. I certainly don't write anything—I don't even come up with the names of the buttons or text field prompts that appear on the screen. But if one of the purposes of technical writing is to tell people how to use the software, a good, intuitive, consistent graphical user interface should fulfill that purpose as much as possible. In this way, I am a technical communicator.

An internship seemed the most logical progression for me at that point. I was twenty-seven, and years in school were the only substantial asset I could put on a résumé. Besides, I didn't know if I'd like the working world, the business suits, and the long days in a little cubicle, stuck in front of a computer. An internship was a short-term means of testing those waters. A month and a half later, I found myself walking into the Prudential II building to begin ten weeks as a technical writer.

The company I worked for makes software for international banking companies. It seemed a strange destination: what does a two-time English major know about banking besides ceramic pigs and ATM cards? I had never even thought about banks needing software, although I expected them to keep track of my money better than I did. In any case, with one other woman, I was to spend the summer in writing field descriptions for the company's new software product. It didn't sound like a lot of fun, but I was eager to work. At least the dress code was casual, and I could wear shorts to work all summer.

Most of the time, I didn't feel much like a technical writer. I had always imagined that technical writers cranked out volumes of text, which product users would keep by their side to explain every step of whatever procedure they were attempting. But what I wrote was never printed on paper, included in a manual, or published in hard copy. I created what was called context help, meaning that users didn't see my work until they clicked on a text field. Then my little three-sentence Help file popped up on the screen for a few seconds. This writing activity seemed strange to me, for I was used to producing paper after paper in graduate school—not to mention generating a 110-page master's thesis just a few months before.

There were thousands of fields to describe. My job consisted of grabbing images of the text field, button, or other object on the screen to be documented, naming that TIFF file, creating a Help file with a relevant name, and writing a short description of what to put in that field or do with that object. While I did create hundreds of these little Help files, I was never able to get too attached to my work. It's a daunting task to document a product that is still under construction. For example, I had written several Help files for a field called Responsibility Unit. A day or two later, while playing with the software, I noticed several other fields called Responsibility Center. Was this a different kind of field, or just a naming inconsistency? My mentor and I spoke with a developer who simply looked at the two screens, blinked, and said, "Oh yeah, it should be Center." That single change made all of my files nearly obsolete, and there was nothing to do except make them all over again. And this type of change happened all the time. I'd spend a couple of hours writing Help files for a certain button, only to go back a week later and find that it didn't exist anymore. It was rather unrealistic to expect the developers to notify me every time they moved a button on a screen or changed the wording of a phrase, but it would have been nice.

Just as the developers were creating their own standards for how the software would look and work, my mentor and I were figuring out how to format our Help files (she had been on the project only a few months before I joined her). Developing a consistent format got a little crazy at times. For example, for a while we were instructing the users to *type* something into a field, but then we'd tell them to *enter* something else. Does one type a name but enter a number? Just when we'd set a standard, we'd encounter an exception to our well-thought-out rule. Our format at one point was to write, "Type the [*x*], then press the Enter key. Your entry appears in the [*x*] field." Easy enough, until we encountered a field called Type and then another called Entry.

Here's another case in point: according to our format, the first sentence of the Help file was to contain a definition of the field or object we were documenting. It made sense to give users a good definition of a term such as *beneficiary bank,* but did we really need to define *city* for them? Though each example may seem trivial, over time they added up and demanded quite a bit of attention, especially when a new format decision could entail trashing a week's worth of work and starting over. This task seemed particularly tough because I was one of only two people documenting an enormous project.

The work was also challenging because I knew very little about the banking processes or terminology I was documenting, and nothing at all about how the software really worked. When I needed help, I had to choose between two resources: the bankers who knew the terms but not how the software worked, or the developers who knew the product but not the banking side of it. It was up to me to tie these branches of knowledge together into one small Help file.

I did occasionally feel a sense of power, though, such as when one developer allowed me to work on the software itself. After the mixup about the Responsibility Center field name, he showed me how to fix such inconsistencies and other problems, including misspellings and misaligned objects, that appeared on the graphical user interface (GUI). His rationale was that as a technical writer, I would naturally have an eye for such errors; I became a sort of visual editor of the way the software appeared on the screen. This editing process involved running a program called Interface Builder and finding the NIB for the screen I needed to work on. (NIB is an acronym for NeXT Interface Builder; NeXT was the operating system we used.) When I first opened a NIB, I was overwhelmed by the array of icons that made up the user interface. It was frightening, like taking the cover off a complex gadget and playing with the guts inside, hoping not to ruin anything. As long as I stuck to working only with certain icons, I could find the word I needed to change or the object I needed to move and get out of the NIB safely. It was scary but also pretty fun to actually alter the way the GUI looked.

At around this time, my internship was coming to a close, and the manager of the documentation team offered me a full-time technical writing position. Putting together three-sentence descriptions about what to type (enter?) in a Receiver's Correspondent field was not my idea of a dream job, but by this time I was pretty settled in my decision to stay out of school for the time being. I accepted, even though my mentor and I were in the middle of changing from writing pop-up Help files to creating an entire Help system using FrontPage, and I hadn't written a single word in about a month and a half.

The next few weeks were quite a blur: my mentor and I were officially switched from the documentation team to the team that was developing this piece of software. We moved downstairs to join the other members of the project. I reported no longer to my mentor but to the man who had built the company website. (He then left the company about two weeks later.) In all the commotion, I deleted an e-mail message that at the time seemed unimportant. However, it eventually would change my career in a way I never imagined.

Sometime during all of these changes, the developer who let me fix spelling mistakes sent me and my mentor a humorous message, asking if either of us wanted to work for him building NIBs—"climbing to new heights in the world of nibdom," or some such thing. I didn't quite understand why he was approaching us with this offer, especially since we were the only technical writers for the entire project. I didn't think he really meant it. Only after my former mentor asked me, a few days later, what I was going to do in response did I give his message any serious consideration. Thank goodness it was still in my Deleted Items folder!

I went to talk with the software developer, who reassured me that the manager of the project really was willing to sacrifice one of only two documentation people to get more of these NIB-things built (an interesting fact in itself!). I knew that my amazing ability to use Word 5.0 hardly made me a computer expert, much less an ideal candidate for a software developer position; this would be totally unfamiliar territory for me. The guilt I felt about abandoning the field of technical writing—and leaving the team with only one writer for the whole project—didn't last too long, however. Besides, building a graphical user interface did look like fun—making buttons that people could click, creating fields into which a user could type information, and, of course, creating field prompts without misspellings and inconsistencies. My job for the past few weeks had consisted of making decisions about our Help system's format and writing next to nothing; in fact, most of the Help files I'd written were already unusable, so creating them had seemed disheartening and unproductive. I also heard that the pay as a developer would be better. I put aside my remaining hesitations and said yes.

Thus I went deeper into the realm of NIB, which is loosely defined as a collection of archived objects, which makes a NIB sound like a library or a file cabinet. Even now I can't come up with a much more specific definition. Building a NIB means putting together all of the pieces, both seen and unseen, of a GUI. When I finish a NIB, a person can run the software and use the screen that I created. Pushing a button with a question mark launches a little pop-up window with a helpful list of values; clicking a Save button saves the data that the user entered; putting letters in a text field meant only for numbers results in an error.

People ask me if I write code, if I'm a programmer of some kind; the common conception of a software developer is someone who writes and speaks another language that makes computers do things. My answer is no: Interface Builder is a program that lets people do object-oriented development. I have a palette of objects that I use to create the interface. I drag a button off the palette and onto a screen, name it, resize it, and literally draw a connection from it to a command to make it do what I want. I didn't have to create the button or write the code behind it that made the four-sided object act like a button that someone can click.

The NIB-building process begins like this: a banker or a developer requests a NIB for a particular purpose—perhaps a screen is needed that will allow a user to capture new customer data and update existing data. All the criteria needed for such a screen have already been created in a database we call the Model. It's up to me to figure out how the screen should look and function and to make it work in the way the bankers and developers intend it to.

Even though many design standards weren't set when I began as a nibbie, several conventions somewhat narrowed the range of my designing power. Most GUI designs fall into just a few categories (the one I described earlier is fairly common), and all screens that allow users to capture and update data look and function very similarly. However, I have been fortunate enough to have a hand in standardizing the finer details of such screens.

Since becoming a nibbie, I've adamantly stuck to my belief that I am no longer a technical writer, per se. I certainly don't write anything—I don't even come up with the names of the buttons or text field prompts that appear on the screen. But if one of the purposes of technical writing is to tell people how to use the software, a good, intuitive, consistent graphical user interface should fulfill that purpose as much as possible. In this way, I am a technical communicator.

A month after I began working as a nibbie, we hired one other person to build NIBs, bringing the total number of nibbies up to four. A month later, the developer who had hired me left the company to work for a bank. Soon after that, the other original NIB builder became our team manager (in fact, he's a former English doctoral student who at one time managed the documentation team). We then hired one more person to build NIBs. In other words, within a few months of becoming a nibbie, I was the second-most-senior person out of four.

I took advantage of this situation, even though the intricacies of Interface Builder were still quite challenging (and frustrating) to me. I quickly put together a list of the few things I knew how to do, partly so that I would remember, and partly to help get the two new people up to speed. This project quickly grew into a several-page "how-to cheat sheet," which I'd update regularly and save in a directory that everyone could access.

I also continued to find inconsistencies in the screens I worked on. For example, the Add/Remove button matrix was usually placed to the right of a browser, but sometimes it was at the bottom of the screen. Some buttons were named Detail, but others said Details. How much space should go between a text field and its prompt? Should a value assistance button be 20 pixels high or 21? It soon became clear that the team needed to make some concrete decisions about such issues, so we held a few meetings and went through each one. And who better to document these new standards than me?

I'm now the NIB team's unofficial technical writer. I'm even working on converting these documents into files that my former mentor, now the project's documentation team manager, can use when updating our old internal website. But technical communication doesn't stop there and isn't limited to me. Even though there are now set standards about exactly where various objects belong on a screen, we made them as intuitive for the user as we could. It's like making a webpage that's helpful and easy to navigate, except that we're quite limited in design; we must achieve the same goal in much more subtle ways.

This points to the importance of visual rhetoric and anticipating how a user will respond to a screen. For example, items placed at the top of the screen appear more important than those placed in a less conspicuous area. Consistency, in both appearance and functionality, is positively invaluable in user interface design. Once users become familiar with how one screen functions, they have an idea about how to use an unfamiliar screen with the same format. If nine screens have a particular button in one place, and the same button is located in a different place on a tenth screen, a user might assume that, or at least wonder if, it serves some other purpose there. And, perhaps most important, inconsistencies damage the users' confidence in and opinion of the product as a whole. These are all issues that graphical user interface developers must deal with vigilantly.

Creating cheat sheets, documenting our standards, and helping redo the internal website have kept me somewhat tied to my technical writing roots, and I don't feel as though I've abandoned my former mentor and her new team. I remember my frustrations with trying to write Help files based on a continually changing user interface, so I go to great lengths to make sure my fellow nibbies and I notify that team whenever we make a change, whether it be an addition, a deletion, or other visual modification that will impact their work. So in a way, I guess I've come full circle—from starting as a student of rhetoric taking technical writing classes with hopes of being gainfully employed, to being a tech writer cranking out three-sentence pop-up Help files, to becoming a GUI developer who relies on visual rhetoric to tell people how to use the product.

Brad Connatser

The Great Pyramid War

Brad Connatser taught fiction writing, technical writing, journalism, and English composition for three years before pursuing a career in technical communication. These days, he manages the publications department for the EPRI PEAC Corporation, an engineering firm in east Tennessee. He is fond of applying cognitive psychology to technical communication, especially to the process of reading technical prose. His research focuses on silent speech, which is the translation of printed words into a speechlike code during silent reading. He publishes articles in technical communication journals and magazines, including *Technical Communication; Journal of Technical Writing and Communication; Reader: Essays in Reader-Orientated Theory, Criticism, and Pedagogy;* and *Intercom Magazine.* He is a senior member of the Society for Technical Communication and a member of the Professional Communication Society of the Institute of Electrical and Electronics Engineers.

Linda's office was austere. Two desks formed an L in the middle. A chair with a crippled back was pushed under the desk nearest to the door. No windows. Nothing on the gray walls. A large, circular stain on the carpet was the only thing of interest in this Spartan setting, and together our eyes drew a bead on this ugliness.

"I'll get maintenance to clean that," I said to Linda. The dark brown stain on the tan carpet smugly suggested a substance that would resist any attempt to remove it. "Somebody must have spilled coffee," I said.

I felt a current of embarrassment rush through me. Why? Why did this spectacle feel like premature failure? I was eager to be worthy of this internship. As the publications manager of an engineering firm, I had conducted two internships in technical communication already, but this intern was a star

> *She had worked on it for about a week, and my expectations were high. I had used the word interest a thousand times, but the story read like a gush of facts rushing down a sluice. There was no opening to speak of, but rather an abrupt sentence jammed with truths: the name of the engineer, the project he was working on, the objective of the project, and so on. Interest had been waylaid by data.*

pupil in a large southern university. Linda was an undergraduate technical communication student who came to her interview loaded for bear: a resplendent résumé, eloquent words of recommendation from her professors, evidence of sincere scholarship, and enthusiasm. Linda was no transplant. She had lived near the university most of her life and spoke the patois of the region. Having been reared in the south myself, I knew that her manner of expression boded well for our relationship.

Her southern origin was evident during our first telephone conversation. "Mr. Connatser—" she insisted on calling me by my surname—"how shall I dress? Is it formal? Or can I wear jeans? Will you supply a computer? When do I get paid? What are the most hours I can work?" She was making all the right noises for an intern, and I answered her questions rather merrily. After all, these were questions I *could* answer. I knew that the tough questions would come later.

On her first day, she dressed in jeans and a white poofy shirt that seemed altogether like an oversized sari, swallowing her whole. I guess that's what they're wearing these days. I was feeling my age. I introduced her to the two desks, the chair, the portable computer, and its keyboard. She had already met the stain.

"This is your computer," I said. "You'll mostly be working with Microsoft Word. You said that you normally use WordPerfect?"

She nodded, or did she shake her head? I couldn't interpret her ambiguous gesture.

"Well," I said, "they're not all that different. As long as you get the concepts behind the GUI, you can master just about any word processor."

I was testing her. I wanted to see how much she knew about common terms of technical communication. GUI, or graphical user interface, was not just a probe into her knowledge. In fact, I assumed that she didn't know the abbreviation. I wanted to see if she asked questions. I wanted her to be the apt pupil here, in this internship, as she was at school, but the probe met with silence. She looked at me through her thick glasses, which magnified her eyes, which perpetuated the appearance of wide-eyed astonishment. Then, she studied her black tennis shoes.

"Do you know what a GUI is?" I heard myself say and immediately regretted it. So much for subtlety.

"Well," Linda said, "yeah, I've heard about it."

I was looking for honesty, but in challenging times, people may paradoxically fudge the truth to preserve their honor. I figured that Linda was doing just that. I had put her on the spot, and it was my responsibility to ease the strain.

"Good," I said clumsily. "Let's go around and meet everybody."

I had gotten my answer. Linda was intimidated by the idea of not knowing things. This affect seemed all too human but not very studentlike. How was she going to develop her professional skills if she didn't ask questions? She was proud, and I would have to learn how to deal with it. Of course, I didn't figure this out from one trick question. This appraisal of Linda came to me slowly, over the course of about a week. I would say something, she would look puzzled and study her shoes, and I would determine what she didn't understand by using the Socratic method of inquiry—lots of indirect questions and reassurance. Students have egos, too.

A week or so into Linda's internship, I gave her a hefty assignment—to write an article about some laboratory work that a company engineer was conducting. When I became the manager of publications, my first charge was to create a newsletter for the supporters of our research, which are mostly electric utilities. Having a master's degree in creative writing, I developed a rather informal, conversational voice for the newsletter. I expected interns to write in the same way, and for the first two interns, that mode of writing was no problem. In fact, I sensed that it was liberating. They rose to the challenge and wrote some very good articles, relishing their freedom to indulge creative language. With their linguistic synapses popping, they approached writing assignments with zeal and ardor, and thereby learned a great deal from the experience.

To instill this philosophy of discourse into Linda's way of thinking, I challenged her to study some of the newsletter issues, especially those issues with articles written by the two previous interns.

"Do you want me to write the article like this?" she said.

I ignored the tinge of incredulity in her voice. "Give it a try," I said. "Read the engineer's report, then write down some questions, and then interview the engineer."

She looked at the ponderous report I had laid on her desk. "You don't have to understand it all," I said. "Try to extract the most interesting elements, and then write some questions that you think might yield some interesting answers."

I wanted Linda to get used to talking to technical types, the subject matter experts who create the data and information that we technical communicators shape and translate into consumable artifacts.

"Is there a deadline?" Linda said.

The truth is, most corporate deadlines play possum. A week's worth of experience in the working world reveals the type of bureaupathic behavior responsible for packing three months of work into a one-month schedule. Deadlines are really just goals. Contrary to the implication of the word *deadline,* nobody perishes when a project exceeds its schedule. Deadlines can be comforting to people who require structure or oppressive to people who wax catatonic under pressure. Linda seemed to be the first type. She nodded when I said, "How about reading this stuff today and interviewing the engineer tomorrow. Then you can write a first draft the next day."

We parted. Linda sat in her crippled chair, and I walked across the hall to my office. A few minutes later, I heard four-letter whispers. Linda was cursing her computer in hushed but intelligible language. I got up and peeked around the corner into her office.

"What's wrong?" I said.

"This floppy drive doesn't work," she said. "It won't eject my disk."

Like some patient with a transplanted organ, her computer was rejecting its own floppy drive. I unfolded a paper clip and inserted it into the emergency-eject hole. The drive reacted to this stimulus by grunting. I tried again, and the disk flew out.

"I'll get the computer guy to look at it," I said. First the stain, now this. "In the meantime, why don't you write the questions out by hand."

I returned to my office. All was quiet. No cursing. She needed time to decompress, no doubt, but I was used to inquisitive interns. I waited for her head to peek around the door. Instead, I heard her office door close and then the opening and closing of the back door to the building, which was just around the corner to my office. Through a window that overlooks the parking lot, I watched her climb into a long yellow car and then drive away. That feeling came over me—the desire to look into the mirror and examine my teeth and the front of my shirt. Or was it something I said?

I coined the term *the complicating why* to describe the penetrating questions posed by my two previous interns. I had to dig deep to explain *why* I write and edit the way I do. Strict and often tiring scrutinies, engendered by the complicating *why,* enabled me to dismiss stale prescriptions and talk about writing and editing in light of reading research, especially research in cognitive psychology. I was beginning to enjoy the benefits of the complicating *why.*

Linda was reticent to ask *why.* She was not laconic by any stretch. A week into her internship, she bubbled with talk about biology, medicine, and horses, but the themes of her conversations rarely led us into subjects closer to the heart of her academic enterprise. I prodded her with points of grammar and mechanics, substance and style. But it wasn't until the second week—while she was crafting that newsletter article—that we had our first substantive and ultimately revealing discussion on technical communication.

I noticed in her writing the tendency to omit commas after introductory elements. At first, I simply advised her that I would prefer that she use a comma after introductory elements. She agreed, but I sensed confusion. I explained: "I know that the rules say you can use the comma after short introductory elements at your discretion, but the writer is too close to the text to decide whether a comma is needed to avoid confusion."

She nodded, or did she shake her head?

I put the following sentence on a large marker board in my office: "In this case only the subject was able to understand the command." With this rather egregious example, I wanted to show her that omitting the comma can cause what linguists call a garden-path sentence, wherein some ambiguous construction compels the reader to make an incorrect assumption about sentence structure. Off the reader goes, down the garden path, headlong into confusion.

"Here," I said, "the sentence has an ambiguous meaning that cannot be resolved without help from the writer. Does the comma go after *case,* which renders one meaning, or after *only,* which renders another? Does *only* modify *In this case* or *the subject*? Only the writer knows, and the writer doesn't come attached to the text."

There was a lot of blinking and nail-biting at this point. Her brow bunched up.

"So," I continued, "go ahead and put the commas in."

Her head gyrated. She adjusted her glasses, breathed deep, and groaned like a slow leak of misery. She wanted to speak, I thought, but was reluctant to disagree. "Go ahead," I said.

"My professor says that writers should *always* leave off the comma," she said.

Always? Always leave off the comma? I was enraptured by this idea. "Why?" I said, suddenly becoming the student. She managed the complicating *why* with enviable confidence.

"Because," she said, "readers recognize the end of an introductory element, so putting a comma there is redundant."

I wondered whether she had understood the example I had just scribbled on the board.

"All right," I said. "I see. But for now, please put a comma after all introductory elements."

Her head moved around a bit, a gesture of assent? Perplexed, we both retreated to our corners. I must think about what just happened. Now, it seems so obvious why this internship so quickly failed to flourish. But as in any James Joyce story, a series of unforeseen human events would suspend my understanding.

One sweltering summer morning, I looked through my office window as Linda pulled her long yellow car into a parking space. I heard the building door creak open and slam shut, then the tinkle of keys, and then the mechanical noise of a lock turning. That was my cue. Her article for the newsletter was due, and I was determined to get it in shape quickly for the next issue. I walked into her office.

"Linda," I said, "how's that article coming?" She was still toting her portable computer, the one with the haywire immune system.

"I can't get the computer to eject my disk," she said. There was no vitriol in her voice, just a calm statement of fact. She was standing in the middle of the stain. "You'll have to read it on the computer," she said.

I took her computer to my office. I had to climb under my desk to find an empty electrical socket. After I booted the computer and opened a file named Article1.doc, I read the article from the dingy computer screen. She had worked on it for about a week, and my expectations were high. I had used the word *interest* a thousand times, but the story read like a gush of facts rushing down a sluice. There was no opening to speak of, but rather an abrupt sentence jammed with truths: the name of the engineer, the project he was working on, the objective of the project, and so on. Interest had been waylaid by data.

In a rare state of repose, I spent an hour in silence developing an encouraging, instructive monologue about the qualities of a good opening, the virtues of a dramatic story structure, the elements of good prose. Certainly she was embarking on a mission of technical communication. Verity was important, but verity wasn't all.

"You're very good with facts," I said to Linda. She was some distance away, sitting cockeyed in her chair. We both folded our arms. The building seemed very cold. "But what you've got here reads like a list. Why don't you pick the most interesting fact in this story and build toward it? Try to create a dramatic arc." She stared at me. Dramatic arc? I demonstrated with a joke.

"Did you read in the paper this morning about the three boys throwing cinder blocks off an overpass?" She had not. "They tied a cinder block to a rope, dropped it on a passing truck, and tried to reel it back. But the cinder block went through the windshield, and the truck pulled one boy's arm off." She winced. "Yeah, they arrested the truck driver."

Now, she is interested, outraged even. "Why?" she said.

"Arm robbery," I said.

She gave me that you-got-me-but-I-don't-appreciate-it look. "Oh," she said.

I said, "Do you see how everything in that little story builds toward the punch line? Upon reflection you can, but while you were caught up in the story you weren't able to see where the words were leading you because the author didn't leave any scaffolding. You didn't see that you were caught up in a dramatic arc. That's what I want you to do. Think about the most important point of the story—a fact or a concept, something you want the reader to think about most—and construct the story around it. When you're done, the editor will make sure you didn't leave any scaffolding standing. You don't want the reader to say, 'I see where this is going.' "

We both sighed, I in relief—I felt that I had done a pretty good job explaining what I wanted—and she because—who knows why? The imposition of a daunting task? The frustration of misunderstanding?

"Well," she said, the dreaded preface to insistence, defiance. Technical writing, Linda explained, should be as economical as possible. The most important information should be near the top, with less important information following. This, she said, was called the inverted pyramid.

Having taught journalism in a small southern college, I was well acquainted with this passé journalistic technique, but I had no idea that it had been adopted by the field of technical communication. The comma quandary did not begin to compare to this loggerhead.

"That's what I've been taught," she added, exerting her full measure of allegiance to that corpus of incontrovertible rules inexorably adhering itself to her long-term memory. Who was I to challenge the wisdom of the professoriate? Thus began the great pyramid war, which would rage on for days, concealed from those around us by a veil of awkward pleasantries. Still, I, the heretic, insisted that she revise the story. She agreed to a one-day turnaround. Okay, then, many thanks. War is hell.

I was excited to see the revision, excited to see what kind of influence, if any, I had exerted on this young communicator, who was now seeming more incorrigible and less impressionable than my first idea of her. I walked to her office. Her head was bent to the keyboard of the portable computer, as if she were pondering its function. I rapped my knuckles on the door frame. She started. I walked to her desk, which was still rather bare. Linda had placed a tin of peppermints next to a sheaf of papers, but the lid was closed.

"Linda, did you repair that story?" I said, then immediately reflected: I didn't say, "Good morning," and that word *repair* was a bit confrontational, loaded with the implication that what she had crafted the day before was not just in need of refinement but was *broken.* For heaven's sake, I thought, absolving myself as guilt surfaced. These interns must be hardened to the real world. What did Abraham Lincoln say? The occasion is piled high with difficulty. That's life beyond the succor of the academy. Every day is a new challenge. She forced a smile as she handed me a floppy disk.

"I hope you can read a PC disk," she said. "It's much easier to work on my computer at home."

I had presented Linda with a less-than-ideal situation, and my expectations of her had been too lofty. The least I could do in the way of recompense was allow her to use her own computer at home. The least I could do.

"Thanks," I said. "I'll look it over."

My word processor translated her file without difficulty. Words popped up on the screen. The title, in bold letters, read "Researchers Study Long-Lead Effect." So far, so good. Then, I read. I rubbed my eyes. I read some more. What she gave me was much less a story than a manifesto, a defense of her position supported by a selection of class notes and textbook quotes.

"Dear Mr. Connatser. Please allow me to explain why I think this article should be written in the inverted-pyramid style." One of her missiles landed in my artillery dump. I was about to reach for a white flag, when I suddenly remembered something. I was her boss. Recalcitrance in the classroom is one thing, but my employer had actually *paid* this intern to explain why she didn't want to follow directions from her supervisor.

"Linda," I said just loud enough for her to hear me across the hall. "Can you come in here?" She appeared in the doorway. "Listen," I said. "I don't want you to think of this place as a word factory where you punch the clock, do your time like an automaton, fill your quota, and leave. But I also don't want you to think of this place as . . . " As what? What was the analog here? An internship is neither fish nor fowl. An intern has one foot in the academy and the other in corporate America. Perhaps this ambiguity was confusing her sense of propriety. "I want you to learn about technical communication, but I also want you to learn to work within the constraints of a hierarchy. So when I say, 'Try to do it this way,' I'd appreciate a sincere effort. Does that make sense?" I asked her that question a lot. Her expressions were hard to fathom, and I had learned to interpret the mostly inscrutable motion of her head as the mere release of contained nervous energy.

"Sure," she said.

Questions came rolling toward me like logs down a hill. These weren't merely complicating; they were confounding, disconcerting *why*s. Linda didn't understand. Could I explain exactly what I was looking for? Exchanges were painful for us both. She seemed exquisitely awkward about repeatedly asking the same questions, and I felt exhausted. Most of my ideas perished by attrition before they reached my lips. My explanations had more angles than a geometry book, but none were lucid enough to convey the idea of a dramatic arc.

"Mr. Connatser, this isn't working," she said. We had failed in the very thing we claimed as our expertise: communication.

Adapting from PC to Macintosh, from the classroom to the office, and from many teachers to one boss was intractable. For Linda, the learning curve was so steep that it became a wall, with a supporting cast of professors on the ground and me atop the wall, exhorting her like a relentless drill sergeant to climb that wall. This polarizing circumstance notwithstanding, winning Linda over seemed insurmountable. It was time to call a truce.

I decided to play to Linda's strengths. She was good with organizing facts and making lists, so I directed her toward a project that had been nagging me for months:

compiling vita information about company employees and organizing that information into employee profiles. Each profile would include current job title and duties, work history, education, professional achievements, and publications. Linda took to this work, despite the computer handicaps and the staleness of her struggle to adapt to a corporate environment.

She worked diligently on this project. As the few weeks of her internship faded, so did her footfalls become more furtive. The hand that turned the key was more deliberate now, so as not to wake the sleeping dragon across the hall. She was putting in time. The money was useful. The project was no challenge, but now she could record some professional experience on her résumé. She must have been disappointed not to have her work appear in one of the company's publications, a piece for her portfolio. In a few days, this would all be over.

Later that year, Linda conducted a student conference on technical communication for her university and the local chapter of the Society for Technical Communication. By all accounts, it was a success. She brought in speakers from all over and handled the affair deftly. At her suggestion, I criticized the flyer that announced this conference, but suggestions for revision were *stet.*

I wondered whether her experience as an intern sullied or attenuated the gratification of success. Oscar Wilde once said, "It's not enough that I succeed; my friends must also fail." I want to believe that Linda transcends that sentiment, that she's not impaling my effigy with mojo pins, that we are both now more humble in our assumptions, and that challenging the assumptions of another can be constructive if it is done with tact, benevolence, and as few stains as possible.

Michael Asay

Glimpse into Reality

After falling in love with writing, Michael Asay decided that he wanted to make it a career. Understanding that creative jobs were few and hard to land, Michael chose the route of practicality—to become an alpha-geek and write documentation for the high-tech industry. He immediately transferred to Utah State University to enroll in their technical writing program. He has worked for four different companies (as an intern, a contractor, and a staffer) since entering the field, writing, editing, and testing documentation for both software and hardware products. He currently works in the R & D division for a Japanese-based telecommunications company in Oregon's Silicon Forest, managing the documentation development for a new product that has source documents exclusively in Japanese. When the right opportunity arises, Michael hopes to return to his creative roots and write copy that invites both laughter and thought.

The sun drifts into my bedroom window as my clock radio goes off—ahh, 7:00 Monday morning. I eagerly throw off my covers and bound down the stairs to a bowl of marshmallow-blasted Froot Loops. After showering, brushing my teeth, and putting on my shorts and T-shirt, I grab my wallet and keys and head out the door just before 8:00. I hop into my Honda hot rod, flip through the stations until I find a good tune, and turn the ignition; I'm off to the races once more.

My mind flashes back to last Thursday's commute home . . . it had taken over an hour to travel the seven miles from IrisSoft to I-5. As I had merged onto the busiest freeway in Oregon, all ahead looked clear; but I was easily fooled, not yet being accustomed to Portland traffic, for I had to slam on my brakes just three miles down the road. What could it be this time? A Jeep Wrangler had flipped and was blocking two of the three lanes; so this was what had kept me from getting to I-5. Over two hours passed before I actually finished my forty-two-mile commute home last Thursday . . . but usually my drive to work is not so bad.

After working on these three different manuals (about twenty pages for each) for three weeks, I feel that they are ready for a technical review (the first in a series of reviews). I print two copies of each manual and quickly scan them for stupid typing errors. . . . I am a little nervous about the review, and waiting alone in the empty conference room does not alleviate the stress or dry the sweaty palms.

I pull into the IrisSoft parking lot just before 9:00, and my spot in the far corner is still available. I love the setting for our medium-size software company:

aged Douglas firs surround the one-level building; other high-tech companies can be seen in the distance. It feels good to be in the heart of the Silicon Forest. I hold my wallet up to the security code box; it reads my card and opens the door for me. I have never worked for a company with such high-tech security gadgets. Inside, I take a left at the first intersection and head toward the kitchen; my nose tells me that the fresh bagels have arrived. I grab two cinnamon-and-sugar-covered bagels and a napkin and head to my office on the other side of the building. I check my mailbox: nothing, as usual. But at least I have a mailbox with *my* name on it. As I walk into my office, the lights flick on; the janitors must have been here last night. I set my bagels down on my desk, turn on the halogen and desk lamps, and turn off the overhead lights. I sit down, reach underneath my desk to turn on my workstation, and push the On button for my gigantic twenty-one-inch monitor. I still remember Pat telling me over the phone that I would have my own office, my own computer, and a twenty-one-inch monitor. I was glad he wasn't there to see my gape-mouth expression: how unprofessional.

After logging onto the network, I check MeetingMaker to see how my schedule will shape up for the day; I have three meetings, not bad at all. During the first weeks of my internship, I had worked alone on assigned projects in my office, but now I actively seek out new opportunities to work with other development teams on various projects. Two of the meetings are in the afternoon and the other is in just an hour. I then check my e-mail: six new messages. The first two are from the CEO back in Virginia, at our company headquarters. The messages give an update on the merger situation, which is going smoothly on schedule.

I recall the day that we were bought out by a partner of ours, BloomWare of California. I had been on the job for three days when Terri, one of the writers on the documentation team, came by my office and told me what had happened. "Now you can put on your résumé that you worked at both IrisSoft and BloomWare." The third message is from Amy, the coordinator for human resources, to remind us of the upcoming company picnic at Blue Lake State Park this coming Sunday. The next message is from Tim of the SWAT team (one of the development teams I work on). He is sending notes from our meeting last Monday. I scroll through the notes, making sure that I had added all the changes to the Readme. My scan brings to attention one area that I had forgotten to document, so I send out an e-mail to Dave, asking if we can meet for a few minutes to discuss this particular documentation issue. The last two messages are from friends from school back in Utah. It is always nice to start off the day by hearing how my friends are doing.

I have thirty minutes before my meeting with Bryan, a member of AnswerLine (technical support). He is the person with the most knowledge of the product that I have been documenting, and I want to meet with him to review the drafts of implementation guides I have been working on these past three weeks. When Pat called me during finals week at school to offer me the summer internship, I asked him if he had any projects lined up for me. He had nothing in mind then, but this is what he had ready for me by the time I arrived.

Some time earlier this year, the interfaces team had released three different interfaces of our main product, HarvestManager (a harvest management tool), but they were released with no documentation. My job was to create implementation guides that would allow users, after installing the interface software, to set up Harvest-Manager configurations and settings and begin using the software in their respective programming environments. I began this process by installing the three programming applications (Visual Basic, Visual C++, and PowerBuilder) onto my workstation. Relying on the brief tutorials that Pat had given me of HarvestManager and through much trial and error, I was able to figure out how to use HarvestManager (and actually understand what I was doing) in all three programming environments. While performing various tasks, I took screen captures of dialog boxes and windows that would confirm for users that they were on the right track; at a later time I would edit the images in Paint Shop Pro and insert them into my document (using Adobe FrameMaker). After working on these three different manuals (about twenty pages for each) for three weeks, I feel that they are ready for a technical review (the first in a series of reviews). I print two copies of each manual and quickly scan them for stupid typing errors. And before I forget, I jot down some questions to ask Bryan about specific procedures that I am still a little foggy on.

Ten o'clock rolls around and I gather my things and head over to Conference Room A, which I had reserved online, on the other side of the building. I am a little nervous about the review, and waiting alone in the empty conference room does not alleviate the stress or dry the sweaty palms. Bryan shows up in a few minutes, happy to see me again. The thing I love about Bryan is how he dresses: today he is wearing green army pants, a blue T-shirt from a softball team he once played on, and an old, contorted baseball cap. His feet are adorned with white boat shoes with no shoelaces. Bryan has to be at work early (by 4:00 in the morning) to answer calls from the East Coast and western Europe.

"Hey, Mike, it's good to see you again. So what do you want me to look at?"

I hand him copies of the three manuals.

"Wow, these look great."

He leafs through them, obviously only glancing at them, not really reading what is on the pages.

"Bryan, if you would look on page twelve in the PowerBuilder manual, I have a question about the procedure for adding users."

He carefully reads through the procedure, not ever glancing up to ponder or question what I had written (he really knows his stuff).

"Actually, Mike, after you initialize the settings of the project, you then need to register the project in source control. Otherwise, you cannot use even the most basic of HarvestManager commands—checking projects in and out from source control."

"Will you explain how to do this?"

"Sure, why don't we go over to your office and work on it there . . . "

Bryan guides me step by step through registering a project. I ask him subsequent questions, and he always has the answer; he even lets me experiment with the answer on the computer, making sure that I really do understand his explanation. By 11:00 A.M., my questions have been exhausted, and Bryan agrees to examine the three implementation guides more carefully. We agree to meet again next Monday morning, and I volunteer to schedule the time in MeetingMaker and to reserve a conference room. He wishes me luck and leaves.

Talking with Bryan has stimulated some new ideas on how to approach the manuals. I open them up in FrameMaker and spend the next hour working on them.

At around noon I walk down to the mini kitchen and fix myself a sandwich. My mom had worried that having me around the house for the summer would drive her food budget sky-high; she was sure glad when I told her about the free lunches I would be eating. A local delicatessen stocks up the refrigerator every Monday with fresh meats, cheeses, tomatoes, lettuce, and other sandwich-worthy materials. I make a turkey and swiss sandwich, grab an apple from the fruit basket, and snatch the last package of Oreos from the upper cupboard. I take my booty to the eating lounge, along with my book, *Pride and Prejudice.* I like to take my mind off technical matters, if only for thirty minutes. Sometimes I gather up the latest copy of the *Oregonian* that is usually scattered all over the counter and see what is happening in the world. No one cares how much time I take for lunch, but I like to get back to my office after thirty minutes so I can go home before it gets too late.

Back in my office, I notice that Dave has responded to my earlier message. He writes that he is pretty busy today, but that he would be happy to let me come by and talk for fifteen minutes or so at around 1:00. This will give me plenty of time to prepare the Readme for our 3:00 meeting. I spend the next half hour brushing up on Defect 6412, the little bug in HarvestManager that keeps Volksbulb, a Germany-based company, from customizing HarvestManager to meet their needs. I know very little about the code of this software, but I still try to read the notes that Dave has left in BloomTracker (an internal IrisSoft tool used to track software bugs and other problems in development). Dave recently implemented a solution, so I want to pick his brain about how he did it—and then put it in the Readme.

At a few minutes past 1:00, I walk to Dave's office.

"Hi, Mike, how are you doing?"

"Pretty good, Dave. Are you glad to be back after that long holiday weekend?"

"You know what, I actually am. So, how exactly can I help you?"

"Well, I'd like to put in the Readme something about how we have fixed Defect 6412, but I'm not quite sure how you went about fixing the bug. Could you tell me what you did to change the code to make it work?"

Dave's face lights up, and a big smile takes over.

"Okay, I'll start from the beginning. Our customers at one German company, Volksbulb, have very strict office procedures. Everyone arrives at work at the same time, everyone turns on their workstations at the same time, and everyone opens HarvestManager simultaneously. However, the HM registry cannot handle these simultaneous openings. It creates all sorts of problems. So what I did was . . . "

Dave explained the technical details while I made notes. He concluded, "And now when workers at VolksBulb open HM, they will no longer have this problem."

"All right, Dave, that makes so much more sense."

"As you know, you get to decipher from this what to put in the Readme. Our audience does not need to hear all of what I just said."

"You know, that is probably one of the most important things I have learned so far this summer—to figure out what my audience needs to hear."

Dave asks about how my internship is going and my impressions so far of IrisSoft.

"After you graduate next semester, if IrisSoft were to offer you a job, what would your answer be?"

Good question, Dave, good question.

"That is a tough question, Dave. So far—though I haven't been here very long—I have really enjoyed the projects I am working on, and even more, I enjoy working with everyone here. I've yet to make any enemies or find something wrong with a coworker. I would probably accept a job here after I graduate."

Another developer, an unfamiliar face to me, drops by Dave's office, and they start talking about national politics and at times insert a joke that gets all of us laughing. It's now approaching 2:00, and I'm still chatting with Dave. I soon go back to my office and open up the Readme. I go through my notes, find the most imperative information from Dave, and quickly compose two short paragraphs for the Readme. I feel satisfied that what I have written is good.

I have about an hour before my SWAT team meeting and am prepared to present my work there. Then Pat comes into my office with Starbucks coupons.

"Terri, Lisa, and I are going down to Starbucks. Phyllis [our documentation manager] gave us some gift certificates. Wanna come with us?"

How could I refuse? It would be nice to go for a walk and get a cool drink on such a hot day. Starbucks is across the street from IrisSoft. We have to cross a busy street to get there, but that is no problem when you're walking with a pregnant woman. "Hey guys, these cars will stop for me," Lisa says. Her baby is due in just a few weeks. Cars do stop for us, but much to my surprise, no one seems to mind; no one honks a horn in frustration.

It is a relief to be in the air-conditioned café. Despite my years spent laboring on a farm during the summers, my toleration for heat is becoming weaker and weaker. Inside, Terri lets out an "ohh" and an "ahh" for a new, hip drink: tea and fruit juice combined. I prefer a vanilla-flavored Italian soda. With cool drinks in hand, we go outside to sit in the shade. Lisa talks about her pregnancy, and Pat discusses his upcoming trip to Chicago to see his brother and go to a Cubs game.

"So, Mike, I've got the big question for you: how is your internship going so far?" Terri asks.

"Well, my favorite part of the job so far is learning about the tools I'm using and the products I'm writing about. The people here are great, too. Sometimes in team meetings—especially when I'm the only writer there—they talk about stuff that I haven't got a clue about. But they never seem to mind when I ask them to explain. I've learned a ton during these gatherings."

We get back to IrisSoft just before my meeting, and I print a copy of the updated Readme. Conference Room D is just down the hall, and I am the first one there. The other team members soon file in: Michael and Terri from product management, Tamara from quality assurance, Dean from AnswerLine, and Tim and Dave from development. The SWAT team is the maintenance team for HarvestManager, and we continuously fix little bugs that irritate users with special needs—like VolksBulb. Dave, the team leader, hands out an agenda that he whipped out on his word processor, and we begin discussing his and Tim's progress in fixing bugs. They set goals for making the next fixes. Tamara and Dean discuss the test cases that they have designed to confirm that previous fixes really did solve problems. And Terri and Michael make sure that we set build dates that will allow IrisSoft to maximize profits and customer satisfaction. And I, the lone writer on the team, give updates on my progress with the Readme.

Before we break, Terri pulls me aside. "Michael, by our next major release we want to have the installation guides updated. Has Phyllis talked with you about this yet?" No, she had not. "I think Phyllis can fill you in on what needs to be updated."

I stop by Phyllis's office—just across the hall from me—to ask her about the guides. "Oh, yeah: Wendy in AnswerLine has a marked-up copy of the UNIX guide. You might want to meet with her; she seems to know that guide inside and out."

I send Wendy a quick e-mail, and she seems happy to meet tomorrow morning. Another assignment, more new things to learn: I love this job.

Pat pops in to see how I am doing. "Everything coming all right for you?"

"Yeah, sure. I just took on another assignment. I'm going to edit both the UNIX and Windows installation guides for HarvestManager."

"Sounds great. How did your meeting go with Bryan this morning?"

"It went really well. He showed me some things that I didn't quite understand, and he is going to look at the manuals and give me some feedback next week."

"Excellent. Sounds like you have the project under control. I don't know if Phyllis has told you this, but we're paying you to work on different projects, not based on how many hours you spend in the office. I know the group that pays your checks wants you to write in the hours that you are at work, but I don't care at all if you want to go home early when you finish all your work. Just mark eight hours each day, even if you go home before that. We're concerned with the job you do, not the number of hours you work. So you can go home now if you want."

And so I do. I shut down my workstation and turn off the lights. I grab a Surge from the refrigerator on my way out. I take the top off my Honda and enjoy my fight through traffic, singing along with the tunes of the radio, not at all worried or concerned with what challenges may face me at IrisSoft tomorrow. My mind and body are now free to do what they please.

Kendra Potts

My Entry-Level Life

Kendra Potts is a 1996 summa cum laude graduate of Illinois State University with a B.A. in English. She lives in Aurora, Illinois, with her husband, Brian, and is a technical writing consultant for an e-business company in Chicago. Her inspiration for "My Entry-Level Life" was not obtained from her current employer, but from her first experience as a technical writer in corporate America.

Does That Say "Sex"?

Yes, sex. That is the one word that catches my eye as I sift through the first file drawer crammed with manila folders overflowing with paper. I pull out the sheet and discover that it is part of our annual sexual harassment seminar booklet, describing how to avoid sexually offending coworkers. ("Keep your hands to yourself and don't say anything you wouldn't want your mother to hear" is the gist of the message, although it gets a little more complicated than that.)

Hello, and welcome to my world (which is currently going through a bit of housecleaning, so please excuse the mess). I apologize for deceiving you. This vignette really has nothing to do with sex. I confess to using cheap shock value to lure unsuspecting readers into my web. But since you're here, stay awhile and explore my realm of existence. Watch your step.

Cubed

I take comfort in the three marbled-gray, slightly padded walls that contain my workspace, where images of my fiancé and cat stare back at me. The pictures are partially hidden by interoffice e-mails of staff meetings, the holiday pig-out notice, and the staff newsletter (written by yours truly). It's the annual "clean your cube and throw out all your crap" week, so we get the privilege of wearing casual attire (jeans and sneakers) while we sort through the

Twenty-five nameless, faceless programmers in a company across the country were contracted to create this software for us. We can't even talk to these developers when we have questions for them. Instead, we have to go through the proper channels, namely, two liaisons who are the only ones in the corporation who know how the product was supposed to work, how it actually does work, and what is happening to make it work as it should. But to this day, it still doesn't work, and we as technical writers still know very little about either the product or creating online documentation.

paperwork accumulated throughout the past year, 70 percent of which, the memo states, is really useless and can be recycled (hence the large yellow plastic recycling barrels in every aisle). Mouse in one hand, and a globe-shaped stress ball in the other, I slowly scroll through the lone e-mail I received today, savoring every word. Not that it was a juicy piece of mail, but the holidays make it desolate around here, and any correspondence is welcome.

My cubed life is rather solitary, and that's the way I like it—however, not quite as barren as it gets toward the end of December. Not that I'm a corporate recluse. I just like writing more than public speaking, or so I tell everyone who asks me why I didn't become a teacher since I have an English degree—as if by declaring English as a major, teaching was the only acceptable alternative. I love to write. But that's not all I do. I write, research, write, edit, write, proof, write, test, write, rewrite, proof, proof, proof, and print. My phone rarely rings, which is fine, because unexpected phone calls make me nervous (if it's not in the documentation, then I don't know—call technical support).

It's really quiet today—lots of people on vacation, not many phones ringing or keyboards clicking. All I hear is the steady hum of white noise.

"Oh, I didn't know you were here today!" It's Bill, one of the technical support employees.

"Yeah, you never know when someone will need emergency documentation!" I quip.

"Yeah," he replies, "as if anyone needs anything typed today!"

I stare at the back of his checked flannel shirt as he walks down the aisle. Then the realization dawns on me: he has no clue what I do here. Now Bill's a nice guy; he certainly wasn't being mean—although to call a technical writer a typist is about as low as it gets—however, the ugly phrase "glorified typist" has passed by my ears more than once. But not many people know exactly what I do—even the people I work with.

Identity

Taking down my "Fancy Felines" 1998 calendar, I discover I'd accidentally pinned it over my graduation picture. There I was, beaming between my proud parents.

For most of my life, my identity was "student." I was categorized by a hyphenated nine-digit number for many years until I received a pricey piece of paper that enabled me to change my identity from "student" to "technical writer." Even though my new identity cost over $12,000 (plus books, housing, and two years' worth of Papa John's garlic bread sticks), that doesn't keep others from trying to recast me according to their assumptions.

I take down the graduation photo and put it in my top desk drawer. Upon further investigation, I discover another picture, long buried: my corporate ID, taken the day I started working here . . .

The Baby

"Here's the new baby of the department!" I heard someone exclaim with a little too much glee as I walked down the gray and white carpeted aisle. I was glad that the second floor had this carpet rather than the vertigo-inducing checked pattern on the first floor.

I turned to see that a mousy-haired girl, not much older than me, had uttered the expletive. I knew that there's an awkward adjustment period in any new job, and throughout the course of the day, the label "the baby" stuck with me like a burr on a wool sweater. Once you think you've gotten rid of it, it pops up somewhere else.

Of course I was young, as many people are when they graduate college and get their first job. I accepted my role as the peon at the bottom of the corporate totem pole. I cut my long, college-length hair, much to my fiancé's dismay, into the "corporate bob" (à la *Friends* style). I bought a Franklin Planner to better organize my projects, what little ones I had to start with. I traded my tried-and-true, tattered and worn, slightly moldy leather backpack for a shoulder-strap briefcase. I was given business cards, the true emblem of postcollegiate success, which I proudly passed out to friends and family and deposited at restaurants in big fishbowls to win a free lunch. But despite these markers of professional achievement, I still bore the stigma of "the baby," the newcomer, the one who, adorned in a cap and gown just two months earlier, had known everything. But once that cap and gown were tossed into the dry cleaning, I reverted back to square one on a different playing board. I traded in the cap and gown for a business suit and, therefore, traded in knowledge at one level for ignorance on a higher level. I once again knew nothing.

But I had learned to play the beginner game once, and I had won. So I just had to do it again. The key, as it was before, would be hard work. That's one rule that doesn't change, even if the playing field does. I knew about the hard work. Other things threw me for a loop.

So I gave "mousy-hair" a good-natured smile and deposited my coat in my new cube. My boss, who had walked me through the department and introduced me to about one hundred people, had finally arrived at "our little family," as she referred to it. More commonly known to the rest of the department as "the doc group," "the tech writers," "the education team," or "the marketing and education team," our family suffered from an acute identity crisis at least once every seven months, constantly struggling to find the phrase that best identifies who we are and what our function is. I'll just call us the family.

Let me also draw you a map of the department within the corporation. The department is IT (information technology). That's the fancy industry term for computers and software. In the corporation there are corporate IT and retail IT. Our department is RIT (retail information technology), formerly RCS (retail computer systems), formerly RS (retail systems). There's that identity crisis. So anyway, RIT consists of support, programming, documentation, and sales. Sales consists of consultants who not only sell our products but also use the documentation we write to train people on the software. There is one bigwig for our department, a manager for each of the four facets of the department, and then about one supervisor for every eight people or so. So there you have the extended family's family tree.

The Family

Like many families of the 1990s, the members of my little group were a dysfunctional bunch. The family totem pole went like this: three documentation writers, three documentation specialists, one senior specialist, one supervisor, one manager, one bigwig, and so on up the food chain. In my two years in the family, we've had two runaways, two adoptions—which resulted in one more runaway, and one "bad apple" we had to toss back. So minus the management, we now have a core group of five, two of whom work at home and are electronically wired into the rest of the family.

Getting to know the members of the family was interesting. They are all women, except for the senior specialist and the bigwig. The age range is between twenty-three and forty; some are single, some are married, some have children, some don't, one is divorced, and I am engaged. It is quite a potpourri. Everyone has their niche in the family too—a flair for creative marketing and design, detailed editing and proofing, writing, technical computer hardware and software functionality, or testing and quality assurance. No one has the same educational background either. I was the only one who had gone to college and graduated knowing that technical writing was what I wanted to do. In the group there is a marketing degree, a communications degree, a public relations degree, a business administration degree, and one other English degree, to name a few. All of them sort of fell into this line of work—I was the only one who went looking for it. Everyone's skills have come in handy because we don't just write documentation: we have to research it, test it, proof it, edit it, and market it. Being the newest member of the family, I had to find out where my niche would be.

I enjoyed proofing more than testing, and writing more than proofing. I was given a little of everything to do, but to this day, I still can't see where my niche exactly is. Still, looking at my desk drawers and files filled with papers, I realize I have accomplished a lot in the past two years.

Getting to know how the family operates as a team wasn't nearly as hard as figuring out how the family operates on a personal level. Immediately I tried to figure out who was nice, who could use a little help in the personality department, whom I could consider a friend, and whom to avoid. Everyone in the family seemed nice—at first. But as the wise Dilbert once said, "Everybody is somebody else's weirdo." How true.

Sarah, a.k.a. Satan

Okay, can you guess who my "weirdo" was? Yes, past tense, *was.* She was one of the runaways, not to my dismay. Sarah was the prodigy, the brain child, the overachiever, the workaholic, the golden child who felt that the department wasn't utilizing her amazing abilities to their fullest potential and that the family was keeping her down. Sarah was also "mousy-hair." Only a year and a half older than me, she never let me forget how I was the youngest of the group. She was a smart person who wanted everyone to know she was smart; but to show her intelligence, she had to make everyone else feel stupid. She was never outwardly mean, but she hurled all sorts of subtle barbs at least once a day. For example, if I'd say that *Sleepless in Seattle* was one of my favorite movies, she'd later bring up how she thought *Sleepless in Seattle* was the stupidest movie ever, and only mindless people liked it. Another ringer was her belief that if you didn't completely pay your own way through college (which she did and knew that I didn't), then you were a big, mooching leech, sucking all your poor parents' hard-earned money. So on a personal level, I tried to avoid Sarah.

On an employee level, working with Sarah could be rather ironic. She did a very thorough job in her assignments, yet when I asked her questions, I got answers that made no sense. She was a technical communicator who lacked the social skills to communicate with real people. I had a hard time understanding why she could do so well on paper and yet, when confronted personally, cause more confusion than clarity. Perhaps by trying to make herself look better than others, she ended up speaking in thesaurus-circles rather than in clear, understandable explanations. Thankfully, she found employment elsewhere, and my days with the family have since been much more enjoyable.

My Boss

Edits from scanning documentation dated September '97 . . . garbage. QA results from the 4.9 update . . . done . . . garbage. I'm quickly filling up the recycling bin when a message pops up on my screen. It's from my boss, Lynda. The message reads, "Have a very merry Christmas!" I consider myself lucky to have the supervisor I do. I learned quickly that your boss can really make or break how you feel about your career, the company, and your abilities.

When I came in for my interview, my boss and I had that cosmic "click" that doesn't happen too often. We see eye to eye on most work and personal issues. It's easy to relate to her and talk to her. I like that. She's a good listener who really cares about what you have to say. Others in the department aren't as fortunate. Sometimes the intentions of an evil boss aren't always apparent—right away. Take, for example, Jack, the support manager.

You Don't Know Jack

"You look really good today!" a male baritone boomed as I walked into my cube. I turned to see Jack, one of the support supervisors striding his intimidating six-foot-something frame up to my cube. I smiled but could feel my face starting to turn turnip. My fair complexion could light up like Rudolph's nose in a matter of seconds. I didn't take compliments from almost-strangers very well, and any uncomfortable situation could make my pasty December flesh turn crimson.

"I really admire tall women who aren't afraid to wear heels." (Smile and nod. Smile and nod. Pretend that it's ten below in here, and there are icicles hanging off your monitor.) Fortunately the comments, and Jack, were passing, so I was spared the agony of stammering through a reply. What the heck did he mean by the whole high-heels thing anyway? I was wearing Easy Spirit black pumps with one-inch heels, not exactly the red stilettos worn by underwear models. I am always careful about the way I dress at the corporation. It's an old, conservative company. Even though the attire was business casual, I had acquired quite a few suits, so I often wore a blazer with a skirt or slacks and pumps. Most people sported khakis and a polo, but being in my first job, I felt that I'd be mistaken for the file clerk if I didn't dress up a little—at least until people knew me.

So the weeks went by, and since my cube was along the main runway of the department, close to the coffee room and bathrooms, quite a few passersby stopped by to chat or say hello. Jack was the most frequent passerby, and his favorite comments were "Don't slouch!" and "Sit up straight!" For, as he told me, "It's horrible when people slouch—especially tall people!" Now I'm no posture queen—I've always been vertically challenged when confined to a chair for extended periods, but having someone point it out continually was starting to bug me.

It wasn't until I mentioned Jack's comments to Karen, fellow writer and warehouse of office gossip, that the floodgates opened. Now, I admit it—just like most of the men and women I know, I am interested in the personal aspects of other people's lives (yes, I'm talking about gossip). I try to be good and not believe everything I hear, especially coming from Karen, who can turn an ordinary event into *Days of Our Lives* meets *Melrose Place*. Her depictions of events have more lust, malice, and backstabbing than any soap on TV. So when Karen divulged the details of how, years ago, Jack was accused of sexual harassment, the details were gory and ornately embellished. Jack was also rumored to have had an affair with someone in the department, both Jack and that someone being married with children. This someone no longer worked for the corporation. Jack was also a notorious skirt-chaser, ogler, and self-absorbed pig, according to Karen.

Weeks later, in conversation with one of Jack's long-time employees, I discovered evidence of more politically incorrect behavior. Jack used to give neck and shoulder rubs—back in the days when you didn't get sued for touching another human being—to "tense" coworkers. However, lately Jack's behavior had improved, perhaps due to the success of our annual sexual harassment seminars. Or perhaps he no longer gets the response he is hoping for from young women like me. The world may never know.

I'm now starting on drawer 3. I toss out two Halloween Snickers wrappers, a cheap chocolate Valentine's Day heart, and a peanut butter foil-wrapped Santa. These little sugar rushes are provided by our department secretary, who keeps the employees just a little wired and the dental industry of Chicago booming with all the candy she buys in bulk from Sam's Club. She's quite the character—I realized this the first time I met her!

Sweetie

"OOOOOOOHH! That pastrami is just going right through me!" This was *way* more information than I ever needed to know from "Sweetie," the staff-created nickname for Elsa, the department secretary. "Hon, be a dear and watch the phones for a minute!" As I stood by the fax machine, she whizzed by me as fast as her sixty-something-year-old legs would take her.

"Now I bet that's a nice vision for you right after lunch," a bodiless voice quipped from somewhere nearby. I turned to see Kathy coming around the cube wall.

"You heard?" I asked.

"Who doesn't?" she replied. "She announces every bodily function and ailment to everyone in a two-mile radius!"

I smiled as Kathy went by, but then remembered why I was standing there. Bewildered, I gazed at the twenty or so phone lines at Elsa's desk. Granted, I had put in time doing secretarial work for temp agencies to help pay for college, but didn't four years and that expensive piece of paper mean that I never had to do this again? Apparently not.

Every phone system I've seen is different from the next. I began to sweat (a little). Why couldn't the calls just go to voice mail? Did I *have* to stand here? This wasn't in my job description. What if someone called and asked me a question? I didn't know anything; I only worked here. What if I had to transfer someone? How was that done? Where was the list of department names and extensions? What if a bigwig called? What if I accidentally disconnected that person? What if I was later tracked down and . . .

"Okay, Hon, I'm back! Thanks a bundle!" I smiled wanly and grabbed my fax along with a handful of M&Ms from the red plastic bowl next to the "I am the boss" plaque on Elsa's cube wall. A chocolate fix was definitely in order.

At the back of drawer 4, I find a stack of Jen's edits . . . recycle material. It didn't take long to find out that Elsa wasn't the only interesting character in the department. Jen, one of the work-at-home writers whose niche was proofing and editing, was much different in person than in my e-mail correspondence.

Jen

"HiSandra!SorryAboutTheMessyHandwriting!HereAreYourEditsBack!IfYouHave AnyQuestionsJustCall.I'llBeHomeByOne.GotToRunAndSeeBobAboutRadioFrequencyChanges.They'reAlwaysChangingThingsAndI'mAlwaysTheLastToKnow! SeeYouLater!Bye!"

That was how face-to-face conversations usually went with Jen, our family's speed talker (conversations were usually over before you realized they had begun) and one of the "lifers" of the department. She had worked her way up from mail clerk to secretary to bigwig's secretary, only to quit work to raise her son. When her son reached school age, she came back and worked her way up *again* from secretary to documentation writer/editor. She had achieved the ten years of employment/ four weeks of vacation marker.

Jen usually starts her sentences with "Sorry . . . " or "This is a stupid question . . . " when she has nothing to be sorry for, and her questions are never trite or stupid. But her lack of higher education has left her with self-doubt and a lack of self-esteem. My analysis: a good-hearted woman but a devil with a red pen. For example, one printer manual included instructions to "plug the male-ended cable into the female adapter port." Jen thought that the wording sounded too sexual and should be changed to omit the gender references. She once told me her theory on editing: "If I don't make a lot of changes, then I don't feel that I'm doing my job." So she sometimes makes changes to the wording that don't necessarily improve it. On the positive side, she usually softens the blow by returning edits with a plate of chocolate chip cookies or other home-baked goodies.

The Grind

No, you'll find no scantily clad MTV dancers here. When I say The Grind, I mean the daily grind, the work pattern that makes up my life. As I dig up another stack of papers in my desk, I find a list of project-scope sheets. Definitely garbage. They may help organize a project, but the deadline portion is a complete joke! I'm given assignments with deadlines, but as I discovered early on, the deadlines are nothing like college assignment due dates. In college, a tornado would have to roll in and suck up the campus in order for a due date to be revised. However, in my niche of the corporate world, I have yet to see one deadline met. Why? Good question. Here's my guess at an answer.

Though the projects are assigned deadlines, many aspects of the job are contingent on tasks done by people in the other areas of the department. Say I'm documenting updates to the accounts payable software. I have to research the changes (and things change every day, so to keep up is definitely a challenge, and in the process of researching the changes, I have to learn the accounts payable system well enough to make logical documentation updates), input the changes to the existing documentation (and if the documentation files no longer exist, I must retype the whole document), have someone from both support and programming look it over (which can take a while because they don't make it a priority), have someone test the procedures, and have someone proof the changes. Then I finalize the work. Sounds cut and dried, right? Not really. Have you ever tried talking to a programmer?

The programmers I've met here don't seem to process information the same way as the rest of the human race. Maybe they love computers more than human contact or are affected by the isolation of their jobs. Whatever the case, it's easy to see why the role of technical writer has evolved. We are their translators. We take their cryptic computer jargon and turn it into everyday language the target audience can understand.

The challenge is understanding the programmers well enough to translate their jargon. There are many roadblocks to communication. E-mail is easily ignored. Say

I call the programmer to arrange a ten o'clock meeting. At ten, I make my way to his or her cube to find it . . . empty. I sit and wait a few minutes, beginning to feel silly. A few more minutes pass, and there is no sign of the programmer. I realize that I've been ditched.

Sometimes a sneak attack can do the trick. Or so I thought. I leave the programmer's cube and decide that by 10:40 he should be back. Well, even then, programmers don't roll out the welcome mat and offer you a chair and a frosty beverage. They ignore you, even when you're standing there. Even when you politely say, "Excuse me." I stood there for at least twenty seconds before he even looked at me. I felt like one of those Asian beetles that found its way to the Midwest suburbs of America—unwanted and definitely where I didn't belong.

But why should I be made to feel that way? The programmer knows I need some answers. I try to ask my question. Do I get an answer? Not to the question I asked. I'm not sure exactly what question he has answered. So I try again, this time rephrasing my request as a yes/no question. I get another bizarre explanation of something that seems irrelevant. He turns back to his monitor, apparently satisfied with his responses. I utter a diminutive "Thank you" and return, dejected, to my cube. I'm still shocked by the treatment I have received.

On such an occasion, my supervisor called to see how things were going. I explained my ordeal and she just laughed. "Yup, those are the geek-heads all right. Not all of them are that bad, but you certainly got the worst one." I wondered how politically correct the term *geek-head* was, but refrained from asking. My supervisor suggested typing another e-mail with specific questions, this time copying the message to herself and the programmer's boss. He'd be forced to respond then, and if the answers came back vague, my supervisor advised me to keep probing. "Don't let programmers intimidate you," she said. "You have to be assertive, or they'll walk all over you. Take charge of the conversation, be courteous, be polite, but don't let them brush you off. Working with us is part of their job, so don't think that you're wasting their time—take as much of it as you need."

Her advice helped, and sure enough, with a little name dropping and messages copied to the right people, I got the information I needed. But what a process! And this is only for software documentation. We also document hardware components, for which we talk to members of support. They aren't as hard to communicate with as the programmers, but there were challenges there too. Once I got all the facets of paper documentation down, then came online documentation, a whole new ball game indeed.

Play Ball

I stare at the two-foot-high stack of binders under my cube countertop—my online edits. Those can't be thrown out. I wish I could say that our online project is as enjoyable as a day at the ol' ballpark, but it's not. The whole concept of a Windows-based project came to the department bigwig in a vision. A Dilbert comic pinned to the wall of my cube sums up the whole project: The boss demands documentation for a product that doesn't exist yet. The tech writer is told to make "logical guesses" on how it will operate, so she writes, "If you press any key, your computer will lock up. If you call our Tech Support, we will blame Microsoft."

Our old UNIX-based system had been around for about eleven years. It had grown and matured through software additions, hardware improvements, and documentation. By the time I arrived, the system was well established, as was our method for documenting it. A template kept all the documents looking consistent (whether from accounting, ordering, payroll, and so on), and a standards manual made the text sound as if it came from one writer rather than five.

But for the new system, no one knew how all the pieces would fit together. Our programmers needed to learn Visual Basic (VB), and we needed to learn RoboHelp, RoboHTML, or some form of online documentation software. Because the bigwig thought this would take too long, a company located across the country was contracted to create the software to our specifications. Our programmers were just supposed to take it over and maintain it. But two years later, we still didn't have what we wanted. Because the software wasn't ready, we were strung along on how to document what is (or isn't, or will be) there. But that was only part of our problem. The daily grind was further complicated by the hired help.

Hired Help

Their purple and gray business cards are still thumbtacked to my cube wall. Everything about their little company is purple: the blocks of lined paper they gave us to takes notes on, the folder they gave us to put handouts in—even their hair had a purple tint to it. Two weeks into my career at the corporation I was introduced to the hired help—two renowned, highly paid "big gun" consultants in online communication. They were originally hired to teach us online documentation for a couple of weeks and then let us fly. Well, that was two years ago. Although they're not on site daily, they occasionally come in to train us, call to review assignments, or e-mail changes to us.

These consultants are usually hired to do the work themselves and then move on to the next job. We don't know if they've ever taught people how to do online documentation in the manner in which they do it—a process that seems to work for them but not for the five of us. So there has been a lot of confusion and struggle with our part of the project.

We've made several attempts at online documentation, but the hired help won't let us get past the paper stages because of the instability of the product. The product keeps morphing to assume different functions and appearances. This amorphous entity has caused a seesaw effect in our online education. (Quick! Quick! Quick! The software is ready to go to the alpha store for testing! Go! Go! Go! We need documentation! No! No! No! They changed it all again. Nope! It's not right yet! Oops! They fixed one bug and created fifteen more.) The worst part is that it's not even the corporation's programmers who are doing this. Twenty-five nameless, faceless programmers in a company across the country were contracted to create this software for us. We can't even talk to these developers when we have questions for them. Instead, we have to go through the proper channels, namely, two liaisons who are the only ones in the corporation who know how the product was supposed to work, how it actually does work, and what is happening to make it work as it should. But to this day, it still doesn't work, and we as technical writers still know very little about either the product or creating online documentation. So we plug along, hoping things will change soon, but continue to seesaw between paper and online documentation.

Quittin' Time

One and a half recycle barrels later, I'm amazed to see how many trees I've killed in the past year, and this doesn't even include my weekly "emptying of the recycle bin" ritual. Just think, the more I write, the less of a chance we have of discovering the cure for all kinds of diseases, cures that are locked in the rain forests we chop up for laser printer paper. We'd better get crackin' on that online help!

As you can see, a great deal of work goes into my daily grind, and in spite of the challenges and problems, I've come to enjoy it. My cube-cleaning exercise has shown me how much I've learned in two years, and I wonder what the new year will bring.

Melissa Alton

First Time Out

Melissa Alton graduated from the University of North Carolina at Chapel Hill in May 1996 with a B.A. in secondary English education. She is also a recent graduate of the technical communications master's degree program at North Carolina State University, where she concentrated on technical writing, web design, and human–computer interactions. Melissa currently works at IBM as a professional information developer. She also designs websites for nonprofit organizations.

I was a naively optimistic, headstrong girl when I left for college without a dime to my name. All I had to sustain me through four years of college was a dream . . . a dream to teach. Naive as I was, I believed that college graduation would mark the end of my angst and economic suffering.

Those days, weeks, and months were long. I paid for my undergraduate education by working twenty to thirty hours a week while taking fifteen to eighteen credit hours a semester. Back then I was a secondary English education major by day and a nanny by night. Life was hard. I was always tired. Outside of school and work I had no life; my only extracurricular activity was the Student North Carolina Association of Educators, of which I was chapter president during my senior year. My sustaining dream was to teach high school English after graduating from college. It took me four years at the University of North Carolina at Chapel Hill to earn my degree and teaching certificate. In those four years I went to a grand total of four campus parties, but I graduated with honors and I owed the university just $3,000.

After graduation I applied for teaching jobs and obtained a position at a small high school in a little town in North Carolina. I was one of those people who went to college knowing exactly what I wanted to do with my life. I had always wanted to teach. Yet I taught for only nine weeks. I was shocked by the disparity between my preconceptions of teaching and the reality before me. During the first week of school, a fellow teacher's arm was crushed breaking up a fight. There were seventeen-year-old freshmen and thirteen-year-old mothers in my classes. There were weapons in my classroom. My students didn't need

> *While grading the computer-based tests and providing technical support, I began to notice flaws in the writing of the test questions and in the design of the software itself. I offered suggestions to the owner of the company, and he liked my ideas. He asked me to write new tests, help redesign the software, and write an instruction manual for the test. I didn't know it then, but I was swiftly becoming a technical writer!*

a teacher; they needed a parent, a friend, a counselor, and a disciplinarian. I didn't want to be any of those things; I wanted to be a teacher. I called the parents of my students every night, hoping to gain their support, but most of them offered no help. The majority of them were not high school graduates themselves. One told me that when her kid was in my classroom, he was my problem, and I needed to learn to deal with him myself. When the principal of the school told her that if her son didn't behave in school, he would be suspended, she replied indignantly, "You can't suspend him. I can't be at home to watch him. He's the school's responsibility during the day."

My friends kept telling me to teach at least one complete school year before tossing in the towel. I committed myself to the semester. I thought to myself, "Surely I can last eighteen weeks." I was wrong. I developed an ulcer, I wasn't sleeping at night, I cried all the time, and my paycheck was too small to cover my medical bills. After seven weeks of teaching, I handed in my two weeks' notice. I lasted for one grading period.

Nine weeks after I left Chapel Hill, I packed all my worldly possessions and headed back. I had no money and no idea what to do with my life. What does a recent college graduate do when she discovers that her dream job is a nightmare? Well-meaning friends and mentors gave me suggestions. Some of them thought that I should become a substitute teacher or apply for a teaching job in the Chapel Hill, Raleigh–Durham area, where the students were better behaved and college oriented, but I was too jaded by my previous teaching experience to consider these options. I didn't want to enter a classroom as a teacher ever again. I still wanted to teach people, to help them learn somehow, but not in a traditional classroom setting.

Because I had no money and no job, but still had a lease in another town, I could not afford to pay rent, so I had to stay with friends while I looked for a decent job. A sympathetic former employer offered me a job as assistant manager in a department store; however, the pay was only $6.00 an hour. I could never pay rent and buy groceries with so little income, so I kept searching. I was at a loss as to how to conduct a job search, so I went to a temporary agency. For somebody with a college degree but no experience, there were a lot of secretarial jobs but not much else. Since I was desperate for money and couldn't afford to be too picky, I accepted a job at a small software company for $10 an hour.

My first day at the new job was a disaster. I stuffed envelopes all day long. I went home and cried my eyes out that night. I didn't want to stuff envelopes for a living. I needed to think long and hard about what I wanted to do with my life, but I was clueless as to what options were available to me. What did ex-teachers do?

Desperate for the income, I reluctantly went to work the second day and prepared myself for stuffing envelopes. But I didn't stuff envelopes that day, or ever again. The person I was replacing began to train me on the company's software. I was going to be grading computer- and paper-based tests and providing technical support to clients who didn't understand how to use the software. My job wasn't glamorous, but it wasn't stuffing envelopes either. I left fairly happy that day, and in the months that followed my job duties continued to grow and my position continued to improve.

While grading the computer-based tests and providing technical support, I began to notice flaws in the writing of the test questions and in the design of the software itself. I offered suggestions to the owner of the company, and he liked my ideas. He asked me to write new tests, help redesign the software, and write an instruction manual for the test. I didn't know it then, but I was swiftly becoming a technical writer! Soon I had so much work to do that the owner authorized an assistant for me. I hired a college student to come in several times a week. I wrote an instructional guide so she could follow the procedures that I was establishing for testing the software. We were still very busy so the owner hired another part-time employee to assist me. Within three months, I was not a temporary employee, but a full-time employee with a lofty title—director of certification, information systems—and I had two employees of my own.

Even though this wasn't a career job, I finally felt I had found a profession in which I was succeeding, not just surviving. I was there for only five months, but I learned and accomplished a lot in that short time. I wrote new tests, helped design new software, established and wrote a manual of procedures for testing the software, wrote an instructional manual for the test, and wrote a manual of office procedures for the company. I worked fifty to sixty hours a week and loved every minute of it. As a result of all of the overtime, I was able to save enough money to lease my own apartment in Chapel Hill. Unfortunately, the owner decided he didn't need my services or the services of my assistants once the tests were all rewritten and the new software was done. He fired all three of us. I'm happy to report that the executive director and many other staff members were not happy with the owner's decision to fire us, so they quit, right there on the spot. The company no longer exists.

Sometime during that five-month period, I had lunch with a friend of mine who was a senior in the communications department at North Carolina State University. I told him about my new job and how much I enjoyed the work I was doing. At this time he told me about technical writing. I had never heard the term before, so I asked him, "What is technical writing?" He told me that I was doing technical writing at work every day. He explained that technical writers write software and hardware manuals; instruction manuals for products, services, and processes; online help; text for webpages; advertising items such as pamphlets, brochures, and newsletters; newspaper and magazine articles; and so much more. He told me that technical communication covers business journalism, corporate communications, manuals, proposals, reports, white papers, press releases, grant writing, and so on. "Wow!" I thought to myself, "I am a technical writer. And I love it! I think I've found my new career. I am still teaching people because they are learning through my writing, and I love getting paid to write. Writing has always been a hobby of mine. I will need some kind of training in *technical* writing, though."

"How do you know so much about this?" I asked my friend. He told me that North Carolina State University offers a master's degree in technical communications, and he wrote a story about the program for the school newspaper. I've always wanted to get a master's degree and Ph.D., so after hearing my friend's sales pitch, I decided to talk to some professors in the program. I talked to the dean of the school and professors who taught technical communication courses, and I researched the field on the Internet. I discovered that the state of North Carolina has three colleges that offer a master's degree in technical communication. Of those three schools, North Carolina State University's program was the most appealing to me. Every professor had a Ph.D., most courses were offered at night (so I would be able to work during the day and go to school at night), the curriculum matched my interests, the tuition was reasonable, and the school was only thirty minutes away. I decided to apply.

I had a lot of preparing to do in order to apply to the program. I would need to take the GRE, write an essay explaining why I wanted to be a technical writer, get three recommendations from employers and professors, get an official transcript from the University of North Carolina at Chapel Hill detailing my undergraduate work, and take an introductory technical writing course. I signed up for the GRE and began studying for it. I also started contacting technical writers via the Internet. I wanted to find out as much about technical writing as I possibly could and find a mentor. After completing these tasks, I was happy to hear that North Carolina State University accepted my application to enroll in spring 1997.

After the owner of the small software company had fired me without warning, I had to find a job, and quick. I had a lot of bills to pay, and the money in my savings account was running out quickly. It's difficult to save money when you make $10 an hour and you're paying rent, buying groceries, paying an electric and phone bill, making a car payment, and paying back your school loan. Chapel Hill is an expensive little town! I called the same temporary agency that offered me the job at the software company, and they placed me at another assignment that same day. Unfortunately, they didn't have any technical writing jobs, so I had to settle for a secretarial job. It paid the same as my previous job, but the work was boring. I was miserable again. I decided that I would work there only until I could find a technical writing job.

Six weeks later I had a technical writing job . . . at least it was supposed to be a technical writing job! I was hired to develop and maintain a website for a small company. Unfortunately, I never got to touch the website. Instead, I was a glorified office manager of a library services firm. I worked fifty to sixty hours a week for little more than $20,000. I did research, supervised four full-time employees, wrote monthly reports for the president of the company, took care of billing, and talked to clients whenever they had questions or concerns about our services. Some of this job was technical writing—the research and the monthly reports. I was also lucky enough to write a manual of procedures and design a new brochure, but the job involved a lot of work that was not technical writing: supervising employees, taking care of billing, and visiting client sites. After nine months, I was offered a promotion that would take me even further away from technical writing. The promotion would not be accompanied by a raise; nevertheless, I was told that if I turned down the promotion, I would not have a job at all. Already overworked and underpaid, I started looking for another job.

I interviewed with many companies for several months before accepting an offer. When I started looking for a new job, I was amazed by all the companies that wanted to interview me. I put my résumé on the Society for Technical Communication's local webpage, and phone calls poured in. I received calls from companies that produced computer software and hardware, advertising agencies, universities, manufacturing conglomerates, industrial plants, contracting agencies, financial institutions, the local, state, and federal government, nonprofit organizations, engineering societies, magazines and newspapers, hospitals, and several Fortune 500 and Fortune 1000 companies.

After much thought, I decided to accept an offer at a Fortune 1000 company that produces computer software and hardware. Now I work for a great company and have a terrific boss.

My current job provides me more than double my previous salary, great benefits, job duties that interest me, and a flexible work schedule. I design and develop webpages, perform usability studies, and teach others how to write effectively. With my flexible work schedule, I can work at home one or two days a week and take classes without worrying about whether or not my work will be affected by my studies. I'm learning new skills at work every day, and I have the freedom to pursue my education while feeling fulfilled professionally.

I still have dreams. It's not as though life stops with finding a good job and a satisfying career path. By the time this story goes to press, I will have my master's degree in hand. But I will not stop there. One day I plan to have that Ph.D. Just call me Dr. Melissa.

Alina Rutten

How I Became a Goddess

In 1994 Alina graduated with an honours B.A. in French/English translation and a certificate in technical and professional writing from Glendon College (York University) in Toronto, Ontario, Canada. After a year of searching for employment, she landed her first technical writing job at CADSOFT Corporation, a developer of CAD-based building/design solutions. While at CADSOFT she has been solely responsible for creating user guides, tutorials, training manuals, online help, demos, newsletters, marketing materials, and interface design. Although technical writing is continually challenging, she finds the profession very rewarding, especially with the recognition and respect she has earned from her colleagues at CADSOFT. Alina lives in Guelph, Ontario, with her husband, Ben, and son, Nikolas.

Dressed sharply in my best (and only) Liz Claiborne suit, I tried to breathe deeply as I waited for my potential employer to enter the room. Naturally I was very nervous. Considering that I was a new grad with no experience, it would be hard to convince anyone that I was the right person for the job. I had lost out to others many times over the past year, so my confidence level was shot. However, after so many disappointments, I had a fiery determination to claim this job as mine.

As I was silently giving myself the "I think I can" speech, a solemn, middle-aged man shuffled across the floor and sat down at his seemingly oversized desk. Little did I know that John would be a key player in my destiny as a goddess.

When he started describing the company, a software developer for the residential building industry, I became a bit uneasy. I didn't know much about building houses, and I certainly didn't know much about CAD. Naturally I didn't let this show, but I could see that this opportunity would be very challenging.

As the first round of questions began, I tried to mask my nervousness with an air of confidence and ambition.

Half the battle was simply trying to understand what I was writing about. The program's interface certainly left something to be desired, and my lack of building knowledge was a definite obstacle. Oddly enough, even the programmers didn't have a complete grasp of their creation. To top things off, I had to learn RoboHelp and attempt to produce a written and online document simultaneously.

Rattling off memorized answers and trying to maintain the unbreakable shell that hid the shy girl inside me, I slowly became a shining star in the interview game. After all, I was an interview pro by now.

After the first round of drilling, I was given the Grand Tour—a sure sign I wouldn't be leaving with just a "we'll let you know." My potentially new environment was an embracing one. The friendly faces, personalized workstations, cutting edge technology, and exciting hustle and bustle of a professional work environment were exhilarating. Although unfamiliar, this workplace was where I truly wanted to be.

Feeling unstoppable at this point, I prepared myself for the last stage of the game. After breezing through the final round of questions (and telling only one small white lie), I felt a positive closure to the interviewing ordeal. Sure, I didn't have any experience, but I had the knowledge and self-motivation they were looking for.

As confident as I was feeling, nothing could have prepared me for what happened next. John offered me the job—on the spot. I think I must have stopped breathing at that moment, because when I opened my mouth, nothing came out. I just couldn't believe that someone was finally giving me a chance. When I came back to my senses, I graciously accepted the position as technical documentation writer.

After thanking John yet again, I gathered up my things and walked proudly out the door of my new workplace. Driving home, I couldn't hold back the tears of joy—a welcome replacement for those feelings of depression I had been feeling while unemployed. With a new purpose in life, I felt invigorated.

My excitement didn't diminish over the next few days as I prepared for my first day of work. When I awoke on the big day, my heart started pounding with anxiety. Actually, I was scared. After all, how could an unsure, inexperienced person like me walk into a successful software company and make things fly? This wasn't just another test or assignment for a teacher to review—this was the real thing, and it was all or nothing.

Feeling comfortable in jeans and a T-shirt, I entered the small, one-story building that I would now call "work." My heart still on the verge of exploding, I was led to my desk. At that time it was located in the technical support office, and because it was situated right in front of the door where everyone could see you, it was referred to as the "sucker seat." Fortunately, my young and personable officemates made me feel right at home with humorous comments and "insights" about the company. Glancing at my computer and all the shiny, new office supplies on my desk, I felt like a little kid on her first day of school. Maybe this wouldn't be so bad after all!

As I was figuring out how to start my system, John came in and sat down beside me. He gave me a quick summary of the applications on my system and told me my e-mail address. I also found out I was getting business cards. "Cool," I thought to myself. I even had the liberty of choosing my own title, so I chose *technical documentation specialist.* (One thing I had learned is that a technical writer does more than writing.) Then he handed me four tutorials that covered the basics of the company's software. Although they seemed lengthy and complex at first glance, I was glad that my very first task involved nothing more than learning. Until now, that's what I had been doing my entire life.

With great enthusiasm I cracked open the first tutorial and dove right into the world of CAD. It was a world apart from the object-oriented, word-processing applications I was used to. There was no point-and-click simplicity here. With each passing day I became more and more ill at ease. I didn't think I'd be able to handle the job. I spent every sleepless night wondering if I would make it to the second week.

Somehow, though, I made it. With superb help from my tech support pals, I finished the tutorials several days later. Armed with the fundamentals of the company's software, I was ready for my next challenge—writing documentation.

The existing documentation, a reference manual and tutorials, was written by company programmers. Needless to say, it went against practically every standard of good technical writing I had ever learned. Not only did it lack style, but also it was disorganized, hard to navigate, and full of grammatical errors. My tech writing profs would have probably deemed it garbage. Many of the company clients certainly did. In a way I was glad because I knew I could at least improve what they already had.

It was easy to see the journey before me would be long and arduous (the contract writer they hired earlier who then quit must have thought so too), but I accepted the challenge with confidence. After all, I had been through hell and high water to get this job. I wasn't going to quit without giving it at least one good try.

In a meeting with developers and managers, I brilliantly presented my argument that a user's guide would be far better than the existing reference manual. This argument wasn't difficult to make, considering that these people didn't have a clue what *usability* meant. After reciting all I knew on the subject, they were easily convinced.

With a clear goal in mind, I began analyzing the software and developing an outline for my new masterpiece. It felt good to be able to apply things I had learned in school to my work in the real world. My excitement about applying good technical communication principles seemed to help me forget (at least temporarily) that I didn't know a thing about building houses using CAD software. After several revisions of my outline, I was ready to start writing.

With a blank Word document before me, I waited for the right words to come. I had always found that getting started was one of the hardest parts of technical writing, and that was certainly holding true now. I decided to skip the introductory matter for the time being and jump right into documenting the software's functions. This material didn't seem to require as much creativity, and it was something I could tackle bit by bit.

Over the next few months I slowly, and sometimes painfully, documented each function of the program. Half the battle was simply trying to understand what I was writing about. The program's interface certainly left something to be desired, and my lack of building knowledge was a definite obstacle. Oddly enough, even the programmers didn't have a complete grasp of their creation. To top things off, I had to learn RoboHelp and attempt to produce a written and online document simultaneously.

Until now there had been no online help, so this was another blank slate I had to fill all on my own.

With a deadline looming, I knew I had to get things in shape quickly. It was almost mid-December, and a new release was scheduled for the first week of January—just in time for the big trade show in Houston later that month. Not only did I have to finish writing the manual and editing the tutorials, but also I had to worry about page counts, covers, and bindings (oh, and don't forget the online help!). I also had to provide a document that would output on a Postscript printer. This was something the company had tried to do previously with no success. I just prayed the fonts I had chosen would work. Since the office didn't have a Postscript printer, I wouldn't know until the files actually went to the print house. "Man, what am I going to do if it doesn't work?" I wondered.

My workweeks grew from forty hours to upwards of sixty hours. Working through scheduled Christmas holidays and pulling my hair out along the way, I was wearing myself thin. My document was so large, it wasn't saving. I often wished I could take an ax to my computer and smash it into little bits. This limitation became the perfect opportunity to provide some input on suitable software for the job at hand. After having a talk with some of my fellow STC members, I proposed to the company that we consider purchasing FrameMaker. Lo and behold, it arrived a couple of days later. I was definitely feeling more like a professional at this point. People were listening to me.

The next day I tried to learn FrameMaker as quickly as possible. The online tutorial made this task rather easy, and after a few hours I was already preparing a template for my manual. Next came the task of converting all the work I had done in Word to FrameMaker. Surprisingly, it wasn't as tough as I thought it would be. Much of my anxiety diminished, now that things were on track.

After providing the initial draft of what was by then a monster of a manual, the first proof came back from the print house. I was incredibly excited when I first saw it. I just couldn't believe the book I was holding in my hands was written by none other than *me!* Hundreds, if not thousands, of people would have my "baby" resting on their shelves. This feeling of elation was brief, however.

The fonts worked out well, but the perfect binding I had chosen proved to be a mistake. The book was too big, and pages were falling out. Luckily, no one saw this as my fault, and after exploring our limited options, we decided to split the manual in two and spiral-bind each part. Unfortunately, the print house we were using was a small operation and didn't have the technology to supply a better solution. I was definitely not happy with what we had to do. Not only would it make *me* look bad, but also it would make the company look cheap and unprofessional. Still, I plugged away in hopes of seeing that ray of light at the end of the tunnel.

Last-minute changes from the programming department made the last leg of my journey very difficult. I thought to myself, "Don't they realize the changes they make affect my work? It's not just a matter of changing some text. There are index entries, cross-references, online help . . . " Fortunately, everyone around me (impressed with my skill and stamina thus far) offered the right words of encouragement. The fact that people had faith in my abilities made the rushed chaos a little more bearable.

The big day arrived more quickly than I would have liked. I am a perfectionist by nature, and sending out something in such a rush seemed wrong. However, I breathed a big sigh of relief when the files finally went out the door. The hell was over. I was amazed with the amount of work I had done—I just hoped that the product would be a hit.

During the next few weeks, I played the waiting game. I wasn't sure what kind of feedback my documentation would get, but I remained optimistic. "At least it's better than what we had before," I said to myself.

Word spread quickly that the company was doing exceptionally well at the trade show in Houston. Our success wasn't surprising, considering we finally had a decent-looking package to present. Previously they had shipped disks along with a binder of 8½" × 11" documentation in a plain, brown box.

At our next weekly staff meeting, the main topic of discussion was, of course, the big show. Sales had exceeded expectations, and our impressive new release had made us the dominant presence among the CAD software exhibitors. The excitement in the room grew as sales figures and customer comments were presented. The head of marketing began openly thanking various staff members for their hard work and dedication involving the new release. Then the attention was suddenly on me.

Apparently both new and old customers were absolutely amazed with the new documentation. It actually contained information that they wanted, and it was usable and easy to read. The documentation was not only a big hit, but also it was a critical factor in the software's new ranking as a leading CAD package. I was, of course, blushing at this point. I certainly wasn't the type of person who liked to be the center of attention. I think that's why what happened next shocked me so much.

When I was thanked, everyone shot up out of their seats and began clapping and cheering. My accomplishment seemed out-of-this-world to them, and at this moment I realized that I had become a goddess. Certainly not in the literal sense of the word (don't get the wrong idea!), but I knew I was something special in this small, bustling company. I was on top of the world and loving every minute of it.

Since my day of evolution from a shy, insecure student to a confident, masterful documentation specialist, I've graduated from the sucker seat and taken on various new projects. I still learn as I go, make mistakes and meet challenges along the way, but I also continue to make a positive impact. Just being the company's number one source for correct grammar and spelling is enough to maintain that feeling of uniqueness and importance.

I'm sure I'm really just an average tech writer, but I'll never forget the time I helped boost a software company into the spotlight, possibly saving it from a gloomy future of mediocrity. Not bad for a kid fresh out of school!

Beth Lee

A Job like a Tattoo

Beth Lee began writing while still in high school and wrote fiction and poetry while working on her bachelor of arts in English at the University of Wisconsin–Milwaukee. She then moved to Marquette, Michigan, and studied for her master of arts in writing at Northern Michigan University. While earning her degree, she was a teaching assistant in the English department, teaching freshman composition. She also wrote a weekly book review for *The Wildcat,* Northern's university newspaper. During her summers, she was an adjunct faculty member at Northeast Wisconsin Technical College, where she taught communications classes (including technical writing). Today she is a technical writer in an information services department at a large retail company in Milwaukee. She resides with her husband, Mike, in West Bend. In her spare time, she enjoys reading, boating, and writing. She is in the process of writing her first novel and hopes to finish it during the year 2000.

I liken the first five months of my technical career to the time I got a Kermit the Frog tattoo during my (very) early twenties. I didn't know if I'd like it, it hurt a lot in the beginning, and it's something that will stay with me forever. However, if I knew then what I know now, I'm not sure if I would have done it at all.

Don't get me wrong. I love my job, and I know I have a real talent for it. I think one common misconception about technical writing is that there is no creativity involved. That's not true—my job requires an immense amount of creativity. Some of that creativity involves dealing with difficult people. I was very green when I began. I imagined that I would show up and everyone would be grateful for the new technical writer. Finally! The developers could get their normal work done and leave the documentation to somebody else. It didn't quite work out that way.

I was brand new to the company, to the technical writing profession, and to corporate America. It could have easily been a disastrous combination. I was naive and underqualified. Lucky for me, I was also eager and tenacious, which was good, because I was the only technical writer in my department. There were no other technical writers to turn to for support and guidance.

I showed up on my first day with some reference books and the notion that everyone would accept me with open arms. However, I was not received in the way I had hoped. I learned later that many developers in the department did not want a technical writer; in fact, they had opposed the idea often and loudly before I arrived.

Even as a little girl I wanted to be a writer. Actually, I wanted to be a Pulitzer Prize–winning, knock-the-world-off-its-feet writer. Later, during my undergraduate years, I decided I wanted to write pulp fiction. At this point I was willing to write any fiction that might make me some money. I even started writing a romance novel once, but gave it up when I realized I wrote boring sex scenes. I then entertained the idea of any career in creative writing: magazine shorts, travel writing, book reviews—anything that wouldn't make my English degree seem a waste. I was never eager to become a member of corporate America.

As luck would have it, I was accepted into an English/writing graduate program. Two things happened to me there that would change the course of my career path. One was my thesis director. He was a miserable pain and didn't exactly hide his lack of appreciation for my writing.

The second was that I began to take a real interest in tech writing. Through sheer luck, I landed a part-time instructor position teaching business writing courses for two summers at Northeast Wisconsin Technical College. I discovered that I was really interested in the subtleties of technical and business writing. So I studied the textbook and taught from it carefully. I also ordered more tech writing textbooks and studied those in my spare time (time I would have otherwise wasted writing fiction).

The head of the comp. department in which I was doing my graduate work had studied for her Ph.D. in rhetoric and composition. When she taught a technical writing theory course, I jumped at the chance to sign up. Her enthusiasm and knowledge of technical writing and its evolution really interested me. She told me I was a hard worker and smart. She often appealed to me to consider technical writing as a viable career choice. Many of my fellow students were desperate to teach in college, but the market was saturated. I would have enjoyed teaching, but I had to be realistic. So technical writing became my focus, and I had some hope that I might have talent for it.

Once I graduated, it was full steam ahead for job hunting. I started out only applying for technical writing jobs. Most jobs wanted too much experience for too little money. So, I bided my time and applied for any kind of job I could find—many of them I wasn't qualified for at all. Only recently has technical writing arisen as a major career path, and many such jobs weren't yet advertised. People found out about them mostly by word of mouth.

As soon as I left the interview for the technical writing position I hold now, I knew I wanted the job. It was in the information services department of a large retail company that's been in Milwaukee for over seventy-five years. The company needed a documentation specialist who would perform many tasks: write user manuals and technical documentation, work on the development of an intranet site, write standards and procedures, and sometimes just proofread. The right individual would have some technical writing experience and a knowledge of Office 95 and 97, among some other specified computer knowledge. I had none of these qualifications nor any real-world experience—not to mention limited computer knowledge. They made me an offer anyway and were very supportive as I gained computer knowledge on the job. They obviously saw some potential in me.

I showed up on my first day with some reference books and the notion that everyone would accept me with open arms. However, I was not received in the way

I had hoped. I learned later that many developers in the department did not want a technical writer; in fact, they had opposed the idea often and loudly before I arrived. One developer would be especially difficult.

Corporate America's politics were new to me, and I was naive at first. I still make a conscious effort to ignore these dynamics: if the issue doesn't specifically relate to me, I don't want to hear about it. Politics, gossip, even just the daily griping—I don't want to get involved. And yet I couldn't always escape it. One morning one of my managers asked me to meet with Peter to write a procedure. When I approached him to ask about a good time to meet, he said to me, "You're just wasting your time. I'll meet with you if I have to, but nobody's going to follow that procedure. Come back in an hour if you want," and with that he turned away.

In retrospect, my blindness and disinterest played to my advantage because I wasn't bothered when people wouldn't help me. Even when people told me, "You can write standards, but no one is going to use them" and "No directory structure is going to work," I just smiled, told them I appreciated their advice, and explained that I was merely doing my job. Then I would hound the person until he or she had little choice but to help me.

Politics run deep in most businesses, and emotions run deep as well. I didn't define what I felt as *hostility* until some developers began to trust me and joke about how they hadn't wanted a tech writer, but now they were glad to have someone who could do a good job with limited time. When a deadline was right around the corner and a user manual still needed to be written, most developers knew they could stop by my cubicle and know I'd make the time to write it for them. I've had people in other departments call me to write manuals or technical overviews. Even the help desk has referred people to me to teach them how to use Word and PowerPoint. I was using it so often that I could often diagnose and solve just about any word-processing problem over the phone. Still, not everyone had warmed up to me.

It began with a simple e-mail that Kim, an especially difficult developer, sent me. I was writing standards for saving documentation as well as a two-hundred-page user's manual for one of our computer programs. I had little time for other work, nor did I expect any.

"You have new mail," announced my pop-up menu. "Do you want to read it?" Kim's message read, in part, "There is a committee discussing [certain] procedures. We meet every Thursday from 10:30 until 11:30. I am supposed to write them, but you will write them now."

I stared in amazement at the message. No "please"? No "do you have time?" Not even "thanks"?

Although I was very upset that Kim assumed she could run my schedule, I was also eager to please and impress the developers, so I accepted the assignment. I decided to ignore her rudeness, hoping it would go away. Perhaps Kim's attitude had little to do with me personally. I've even seen her intimidate some managers. In fact, my final showdown with Kim occurred because one of the managers didn't want to confront her himself.

I attended several of the weekly procedural meetings and wrote up some procedures. Kim wanted to read them prior to my dispensing them to the others, so we had a meeting to discuss them. She began, "Well, I know you're used to being in school and adding extraneous material to just fill up space to get the necessary amount of pages. We don't need to do that here."

I smiled but began to doubt my abilities.

She then added, "I brought some of my procedures. I think you should set yours up like them."

Okay, I told myself. Maybe she's right. Maybe it isn't that she just enjoys putting me down. Maybe these are really terrible. Maybe I'm no good at technical writing—and so my agonizing began.

Meanwhile, I finished the user's manual that I had been writing for another developer. He shook my hand, congratulating me upon completing my first big project. "I really like your work," he said.

I began to regain my confidence.

The rest of my confidence reappeared during a long talk with my boss. One benefit of my job is that I receive little or no supervision from my boss, Jim, who is the director of information services. I am one of two people who aren't management but answer to him directly. He was usually out of the office or in meetings and otherwise occupied, making decisions for the department. So we'd try to meet on a somewhat regular basis, and if I had questions he would always clear his schedule to speak with me in a timely manner. I also worked closely with the other managers, as they often arranged for my document projects.

It was a great arrangement for both Jim and me. He's very busy, and I prefer to work autonomously. Most of my projects are referred to me through the director. Managers who assign me work often tell Jim what the job consists of. So, although the director and I are not in constant contact, he usually knows what I am working on. If I'm not sure whether or not he knows, I e-mail him with an update. I also write a monthly status report that he reads.

I happened to be in his office one morning, copying some diagrams a developer had written on his board. He was in a jam and needed me to put the notes and diagrams in Visio Professional for a meeting he had later that day. As I finished my note taking, my boss and I began to talk about my job.

"I hope that this job is what you expected and that you are happy here," Jim said.

"I am, Jim. I really enjoy what I do. I wasn't sure, though, if I was qualified for this job. I really second-guessed myself," I answered.

"Well, we are very glad that you took this job. You have a very good personality for it."

I left his office, feeling good about the work I was doing. And, knowing this, I decided I would not take Kim's nastiness anymore. I just wasn't sure how I would put an end to it.

I was still writing standards and a directory structure for the department. It was an irritation to many of the developers, but it was a necessary evil. One manager asked me to ask the developers how they felt about the directory organization. We had already had a meeting to explain the proposed directory structure. We knew there were some problems with it, but we'd put together a presentation, and I had sent out an e-mail about it so that everyone could review it. I knew I'd get a different opinion (and possibly a lecture on what was wrong with it) from every developer. And I knew my least favorite developer would be the most difficult. So I waited until the end of the day to approach Kim.

"Hi, Kim. I'm just asking everyone if they have any opinions about the directory structure or ideas for improvement," I said lightly. She was still typing on her computer, and I was hoping to escape quickly. She turned slowly, and I knew by the look on her face that escape wouldn't be easy.

Kim blinked and said, "Well, no directory structure is going to work. You're wasting your time."

Already, I was starting to burn. I said, "I understand you feel that way, but Jim and Tom want some directory structure, so I am going to implement some system."

"No one is going to store documents in any kind of structure. You're wasting your time."

I wanted to scream, "You rotten bitch with a bad attitude! No wonder you aren't part of management! You're totally unprofessional!"

Instead, I asked, "Well, if you think our structure is so bad, what do you suggest we do?"

"I suggest you don't bother. You're wasting your time. No one is going to save anything the way you want."

"Kim," I said, "it's not the way that I want. I am under orders from management to come up with a directory structure. I am merely trying to get some input from everyone, including you. I will be sure to relay to management your sentiments." I walked away.

Back at my cubicle, I was still steaming. I was also angry because I suspected the manager had sent me to talk to her knowing what her reaction would be. So I wrote a memo to my boss and to the managers reporting the feedback I had received. I added all the positives, but I also noted: "I received some hostility from certain individuals in the department, although I suspect none of this surprises you." Then I quoted my conversation with Kim.

Several weeks later I sent out an e-mail that described our new directory structure and announced that the manual explaining the structure in detail was available online. Almost immediately, I received a response from Kim:

"I don't understand these standards. I need more information immediately."

I quickly wrote up a two-page explanation and pointed out again that the manual was available.

Kim wrote back, "I don't have time for your fifty-page manual explaining this. If I wanted to bother with all that, I would have already done so. Is there any other place for information? I don't understand this."

I responded that there were very few changes from those explained at the meeting and that any further questions or comments she had could be directed to her boss. I copied him on the e-mail as well. Not long afterward, I saw Kim enter a closed-door meeting with our manager. As much as I wished that the meeting had been called by the manager to tell Kim to shape up, I suspected that Kim was running the meeting, complaining about me.

Soon after that, I began to feel that I had hit my stride and had found my way fully into the company. I developed an internship program in our department and began working on the design and implementation of the company's intranet. I became increasingly busy with requests to do documentation projects for developers. I learned a documentation tool that centralizes most of our documentation, whether I write it or not. I moved my cubicle right into the middle of the developers' area.

When I returned from a meeting one day recently, Kim peeked into my cubicle.

"Could you give Nancy a call? She's having a really hard time taking screen prints. She needs your help. The help desk gave her vague directions that aren't working. Oh, and by the way, if you have time next week, I'd like to sit down with you and go over the revisions for the DOMS documentation you did for Tom. They're fine, but you need to clarify a few things."

I happily called Nancy to help her. I couldn't believe it, but I had weathered the storm with Kim. That, too, felt like a professional accomplishment.

Carol Hoeniges

It's Not Mark Twain's River Anymore

Carol Hoeniges is a writer with the Hile Group in Bloomington, Illinois, a consultancy specializing in performance improvement and workplace documentation. Hoeniges partners with clients in operations and safety departments to develop various kinds of organizational documents, including policy and procedure manuals, system guides, and safety rule books. Before working at the Hile Group, she worked in the banking industry and as a technical writer for a software development company. Hoeniges is a senior member of the Society for Technical Communication and presented a paper on team-based writing process at the 1999 STC Region 7 Conference in Seattle. She holds an M.A. in professional writing from Illinois State University and a B.A. in English and folklore from the University of Illinois.

My knuckles white, I clutched the steering wheel for the entire fifty-minute drive to Peoria. Why was I so nervous this time, I wondered? After all, I'd been through the hiring drill before: résumé, letter acknowledging the résumé, polite phone call to gather more information, cordial phone call to set up the time of our meeting, blah, blah, blah. So far so good. I liked Emma already. So why all the panic? "Well, for one thing, Carol," I said aloud, "you look like a big dork with that cast sticking out below your parochial-schoolgirl plaid skirt." Maybe I'll get sympathy points when I hobble into the reception area. Okay, what do I know about ARB Technology Solutions? They were a software start-up company, owned by three men. ARB developed customized software for industrial applications and wanted to hire another writer to fill a business writing need they had. This need included creating the employee handbook, designing business forms, generating the company newsletter on a regular basis, and so on.

Emma said that ARB was located in a newly remodeled building in the heart of the city's riverfront rehabilitation project. She also mentioned that ARB was moving. As I pulled into the parking lot, I briefly wondered where to.

> *The three stony-faced engineers back there look as if they are plotting a coup. "Let chaos happen," Julie had said. Well, it just might.*

The reception area wasn't as formidable as I had thought it would be. In fact, it included just me and a very young receptionist. I waited, trying to be patient with the discomfort of the cast on my leg and with watching for Emma to appear from behind the security door. As if she had read my mind, she opened the door. I was face to face with a slender woman, probably of my age, full of energy. Her vibrant blue eyes were set off by brown hair. Her smile welcomed me, really made me feel that she wanted me there. It was the start of a beautiful friendship.

Two months later, Emma and I were sitting in adjoining cubicles, plowing through our respective writing projects. So far, I enjoyed the blend of work, particularly when it focused more on business topics. There was something about how the writing impacted people that really struck a chord with me. From past work experiences (many of them negative), I felt very capable of taking the employee perspective to heart and knowing the impact of policies on the individual and the work environment. Even though new to the company, I wanted to advocate for my coworkers. I wasn't just a tech writer who would write anything that anyone wanted. I often struggled with managers and colleagues to find just the right wording and to caution them against the impact of bad policy.

Emma was the only other tech writer on staff. Her position had evolved into an intriguing and challenging blend of writing, project management, software development, and negotiation with the owners of ARB. She was a techie at heart and extremely capable of taking on the programmers and engineers when it came to software design.

I looked over my shoulder to see Emma typing furiously in a Word document. She systematically filled the screen with topics, all organized in a logical hierarchy. How did she do that? I sat in silent amazement, then decided to ask her.

"How do you do that?" I asked, breaking into her feverish typing.

She rolled back in her chair, looking puzzled, and replied, "What do you mean?"

"I mean, you can slam out a user manual faster than I ever could. You open up a blank document, set up headings and subheadings, and fill in the blanks with info you drag out of the programmers. Not to mention the fact that you're the goddess of Word and Doc-to-Help." I went on to tell her how I was even more impressed by her writing ability. She never let her "technicalness" get in the way of her end result: writing a user-friendly, reader-accessible document.

She hesitated, almost embarrassed by the praise. Finally, she said, "I'm amazed by your ability to build a story, to look at things from such a broad perspective while still nailing an idea right on the head. I look over and you're just sitting there in silent, deep thought. I used to think you were spacing out. But then you would start typing nonstop, and *voilà,* you would come up with a full story, thorough but without filler and BS."

She continued, "I think part of my speed on the technical side has to do with the fact that I actually thought about being an engineer. I spent one year in Iowa State's program. But it just didn't do enough for me, so I switched to English for my undergrad, which seemed to complement my music degree better. I focused on technical writing for a master's degree because it kept me linked with the technical side, but studying writing seemed to keep me creative."

I started to consider my own background and how it influenced me. I told Emma about how I realized, in hindsight, that Illinois State's program was very well suited to me. I was required to take not only technical writing classes, but also those involving rhetoric, composition, and literature. While others found the nine hours of literature painful and tedious, I welcomed them as an escape back to my happy place: the home of stories and artistry.

I told Emma how I had been surprised to be fascinated and profoundly affected by composition theories and how the concept of social construction showed me how I was linked to the larger world of community. I started to think about people and how they connect, and considering such relationships had a significant impact on the way I write. How do I live and move among various communities? How do I adapt my role within a specific cultural context? And, more important, if I do change to fit in with different circumstances, how does this affect my writing? Is it better? worse? How do I negotiate and navigate within the community to reach acceptance?

After intently listening to this dissertation, Emma started to talk about our similarities. "Even though my work at Iowa State didn't go as far into English studies, I think that you and I have the same sense of audience. We both take time to consider the people who will use and read our books. If we can talk to them, we listen to what they have to say and use that input in the final design of the manual."

From there we went on to talk about visual design, Edward Tufte, and William Horton. We had both studied design to some degree and were both drawn to the ways that the text and visual components of a document can work together. Emma also had work experience in applying her visual design skills to websites.

We paused, heady from our hour-long conversation about writing and communication, feeling mutually supported and appreciated. Yet over the next few months, this idealism began to fade for me. It didn't seem to get me very far on the software-related products. I quickly saw that although managers at ARB knew they needed tech writers, they didn't really know what to do with us. We were often brought in at the last minute, far too late to make any positive impact on a product. Little did these managers know how we (especially Emma) could help them with usability testing, with GUI design, with plain and simple logic. I, too, could have been particularly helpful in these arenas, ironically, because I didn't favor technology over people.

As I had done so many times before, I rolled my chair to Emma's cubicle.

"Emma, when you're done with that chapter, can we talk in the conference room?" She nodded, looking somewhat like a deer caught in headlights. I closed the door as we entered. The light blue walls seemed darker, the conference room smaller. Emma sat on the corner of the oblong table, making it very difficult to avoid her vivid blue eyes. "I'm quitting ARB," I blurted out with an utter lack of refinement. Tears welled in Emma's eyes. I picked at a spot on the table and then swallowed to remove the prickly lump that had formed in my throat.

"I have a job opportunity with a woman who owns her own consulting company in Normal. She needs a technical writer to work on a safety rule book, one who knows Quark. From talking to her, I think that I'd be better suited to the work ethic and environment that she has created. Her work is focused on helping people learn. Unlike here, she develops documents and training programs collaboratively.

She likes to think outside of the box for solutions." I paused, waiting for Emma to respond. Instead, she looked ill.

The silence deafened me, so I continued, "It is going to be hard for me to leave you. I've enjoyed working with you and becoming friends. I hope the friends part will continue. I just don't think I want to be a part of this business anymore. I feel undervalued because I don't clock as many billable hours as you do. The owners imply that my internal writing projects don't build the company as customer projects do. I'm tired of some programmers' attitudes, of being treated like a glorified secretary. They bore me, really. Not everyone is like that here, but you and I are never truly involved in product development. I want to be someplace where I can make a difference, where my work pays off. If I want to have 'real' value at ARB, or what they think is real, I will have to write software documentation, learn object-oriented programming theory, and immerse myself in GUI design. I would have to become a techie in order to stay there. Deep in my heart, that kind of change feels like losing something, something creative and personal." Again I paused, picking again at that spot.

With a heavy sigh, Emma composed herself. "I can't say I blame you, but this really sucks. I've been discouraged about the company's attitude, but I've always challenged myself to overcome their attitudes about gender and their myopia when it comes to software development. But I'm getting tired of it, too. Brian isn't happy at his work either, so we've been talking about our options. I really wonder what's going to happen here. I've been here long enough to want to stick it out. But on the other hand, this cavalier attitude they have in running the business worries me. Should I get out now or hang on? Without you here, the staying will be painful. But I really do support your decision. It sounds like the best thing for you. The woman you know—she's not hiring two writers, is she?" Emma tried hard to smile, but her heart just wasn't in it.

I had avoided him for the two weeks since I'd given my notice. I really didn't want to do this exit interview with him. And who knows what he would report back to the two other owners. Now was the time to decide if I was going to burn this bridge. The sight of the second owner, Saresh, approaching me broke into my thoughts.

"Are you ready for your exit interview?" he asked. As it was so often, his question was rhetorical. He really didn't care whether I was ready or not, so we proceeded into the conference room.

After speaking briefly about the commute and other choices I'd made for myself professionally, I decided to launch into some of the real problems I saw at ARB. Oddly enough, I was still hoping that he would take some of it to heart and make some changes. The prickly lump was back; I swallowed hard.

"I think that you underappreciate the contribution that your writers could make on your software projects. I'm fully aware that my experience is lacking on the software side; however, Emma's technical knowledge is significant, and she is still underutilized. We both bring user-focused perspectives that can add real value to the way the programmers do their work. But as long as they are allowed to sit in their cubes without contact with the real world, bringing us in at the end of development, the software will continue to be nonintuitive for the user. It won't be the product you want it to be.

"There is a lot of talk around here of teams and teamwork, but it seems to me that it is just a bunch of lip service to an ideal that just isn't happening. Often, my role on the so-called team is to 'fix' something that the programmer already wrote or to do some kind of grammar magic. But Emma and I can bring so much more. We can help with logical thinking, project management, GUI design—I know you think I'm leaving because I'm unhappy and hate the commute. But in reality, it is because I don't want to spend every day frustrated at knowing that I could make a difference if only anyone else besides Emma cared. Until there is some progress in the way projects are managed and until you bring writers like Emma in early on, I'm not sure that things will improve." I trailed off to take stock of what I had just said and wait for Saresh to respond. He was surprisingly calm, looking at me with his caramel brown eyes.

After a weighty pause, he said, "It seems to me that you really just don't want to commute, that this has been hard on you. I wish you would have said something. With all the technology at our disposal, we could have fixed that, set you up at home."

Right, I thought, technology can solve everything.

He continued, "I think you could have learned from Emma more of what we do. It may have made more sense to you. But it sounds as if your new job is better suited for you."

His tone, more than his words, suggested that he meant I was leaving because I wasn't equal to the demands of a technological workplace.

"The pace here is intense," said Saresh. "We have to keep up with technology, to move forward. We don't have time to look backward."

At the risk of not looking forward either, I thought. I knew he was again dismissing what I had said, writing me off as someone who wasn't a techie and who was leaving for personal reasons. I wish I had that wasted breath back.

That was three years ago.

My husband has often said that I have the Midas touch when it comes to finding jobs. I found my new job by doing contract work for Julie while working for ARB. She owns her own consulting group, helping transportation companies build training programs and write documentation related to safety, health, policies, and procedures. When I said yes to the position, I didn't really know what I was getting myself into. I did know one thing for sure: Julie focuses on people; her energy and enthusiasm for the work were what I found intriguing and challenging.

I sit at my new oak desk without a cubicle in sight. Across the room, Julie is engrossed in a conference call with a port captain and a lead trainer for a major U.S. barge line. She is helping them develop a more cohesive and interactive training program, one that opens the lines of communication among vessel crews. I, on the other hand, am staring at the paper strewn across my desk, frustrated and confused. For the first time, I'm the lead writer on a manual for another barge line company, National Marine Transport (NMT). The book covers towboat operations, including policies, procedures, and resource information. We call the book VOM, which stands for *Vessel Operations Manual.* We're using a team approach to writing, and I'm partnering with the project manager, a port captain named Dave. I leave tomorrow for our first team meeting, but I feel as if I'm missing critical information. Even though I've tried, I just can't find answers to the multitude of questions that I have about facilitating this meeting.

Still knee-deep in source materials and questions, I patiently wait for Julie to hang up. In the meantime, my mind starts to wander. Sometimes I almost long for those days when I could just be the introvert at the keyboard, generating text without much contact with the outside world. At least then I needed only to process my own thoughts. But that tactic didn't lend itself well to the consulting role that Julie expects me to develop. I was often so wrapped up internally that I would catch myself not listening. I missed or got only part of the information I really needed in order to do a good job of writing for my reader. Through the last three years, I had watched Julie for clues, seeing how she gathered information and listening to the questions she asked. While learning the writing style of the company and our clients (their tone, their voice, their jargon), I was forced to question how my evolving communication skills and background were compatible with Julie's already mature approach.

"Do you have some time?" I ask. "I need to finish planning my VOM core team meeting." Julie has finally hung up the phone.

"Where did we leave off from yesterday?"

"I was starting to tell you about who is on the team and possible options for the agenda that Dave and I have talked about. Should we start there?"

"Definitely," Julie says. "Tell me who you've got coming to the meeting and why they are there."

"There are going to be around fifteen people, if everyone shows up. From the boats, we've got two captains, two mates, three engineers. The mates know all the deck jobs; that's why we don't have any deckhands. From the office side, Jack, the compliance guy, is coming. So is Don from safety. We've also got two more port captains that work on training programs, in addition to their other work. Will is coming up from the fleet office in Cairo, Illinois. This will help since the fleet and harbor boats have different operations than the main line boats that run the large rivers. Overall, I think that Dave has done a good job of selecting a diverse group. The boat guys are all involved with the safety certification process that helps them become familiar with the manual. I think they will have some great insights.

"But I'm very stressed about my lack of experience in facilitating this kind of meeting. What if the team members resist being involved and jeopardize the whole process? How can I manage dissent and keep the process on a forward-moving track?"

Without letting Julie respond, I forge ahead in my anxiety. "I know that I need to accomplish two things early on in the meeting: I want the guys to realize that this is going to be an interactive, thought-provoking session and not just a sit-there-and-listen meeting. I have to find out what preconceptions they have about the manual and the project. For introductions, I'm going to have them partner with someone they didn't know or know well. After they introduce each other to the group, one of the questions I'll have them ask each other is 'How do you use the VOM? If you don't, what do you know about it?' I'll have them write their answers on flip charts. To go a little deeper, I'm going to post two flip charts, one with the heading 'Positive' and one with the heading 'Negative.' I'll ask them to write down at least one specific detail, a positive and a negative, about what they know or think about the VOM.

"After this, Dave is going to give some background on the revision process so far. I hope that he can help the team understand why they are there and establish his authority as lead at the same time. He is going to keep it brief so that we can start working on the text. And this is where I get really nervous because I don't know what to do next. How will I know?"

Julie ponders, letting my stream of information soak in. "First let me say that I think your facilitation ideas are solid. It shows your team members that their involvement is necessary. As for your stress, I know you won't believe me when I say you're ready, but you are. Sometimes you have to let chaos happen, let ideas flow. In this kind of interactive process, you often take two steps back to move forward one. In the end, the answers surface, with the help of your team members. Plus, you have Dave and Don there to support you. They won't let you down."

"They won't let you down." Julie's words still ring in my head as I move to the back of the NMT conference room at NMT's headquarters in Paducah, Kentucky. The flip charts the team has produced in the first two hours of our meeting fill the walls around me. Gary, the chief engineer, has been grimly silent the whole time, other than occasionally mumbling something to his engineer colleagues. I wish that I had done something to mix up the seating at the tables. Those clusters of three or four at each table don't seem to be helping with whole-group interaction. The three stony-faced engineers back there look as if they are plotting a coup. "Let chaos happen," Julie had said. Well, it just might.

Dave's voice and the group's discussion sound distant. It's now or never. If I don't get Gary involved, our success as a team is in jeopardy. With my VOM in hand, I eye the row of chairs against the wall behind Gary. I meander nonchalantly, looking as if I just need a place to sit down with my big book. It isn't long before Gary mutters something under his breath. Carpe diem! I lean over and start talking to Gary's back. "What did you say, Gary?" I ask quietly.

Gary looks startled on one level, but oddly expectant on another. He hesitates. "I said that I don't think it is a good idea to move the text regarding engine maintenance and fueling to the section after cargo. That's all."

I wait for a pause in the group discussion. "Dave, Gary has some input back here on the sequence of that maintenance info. Gary, can you tell Dave more about it?"

Again, Gary hesitates. Is he going to move on it or not? My heart races. I know that this is a turning point. If I don't get him in here, the meeting will be a wash from the engineering standpoint. Maybe putting him on the spot was a bad idea. What if he totally clams up?

Gary shifts in his padded metal seat. As he starts to speak, I hold my breath, listening intently, feeling my tension begin to melt now that he is participating. He's looking at me for reassurance as he explains his reasoning.

Somehow, I manage to prompt him calmly. "That's a great insight."

He continues, and I can sense that out of the chaos, some order is emerging. What he is saying is rational, clear. The others are nodding; you can hear the jumbled pieces clicking into place. Obviously, Gary is really smart. These towboaters know their business! And big business it is; it isn't Mark Twain's river anymore!

Boy, Julie is on the phone a lot today. I should make some notes so that when I have her attention, I can plow right in, asking her about the VOM plan. It doesn't seem possible that a whole month has passed since the last meeting. Our next one is coming up in a couple of weeks, and Dave and I still don't have a full plan. How are we going to get the team members to tackle the content, to get involved with the writing and revision?

My reverie is broken by the sound of Julie hanging up the phone. "Do you have a few minutes to talk to me on this VOM stuff?" I ask. As she nods her head yes, I continue. "I talked to Dave today about our next core team meeting. We decided that we have to get the team members working the content. So far, they've only reorganized the document. Granted, the work they did was great. I never would have come up with that structure in a million years. I just don't know their jobs well enough. The only drawback is that they have moved around existing material without getting into an assessment of the content. I don't know how to get them engaged with all the company policies and work procedures. They just don't feel empowered or smart enough to do it, even though Dave and I know they are.

"Dave and I want you to consult with us on this, as a teacher of writing process. If you could lead the content review, Dave can work the small groups as content expert. I would be freed up to work the VOM as a writer, without thinking about leading the meeting. What do you think?"

"How do you think you'd use me in the meeting?" Julie inquires, munching on sesame sticks.

In my mind I assemble all the players in that NMT conference room in Paducah, and I begin to explain.

In the midst of the second meeting, it becomes obvious why Julie is needed and how she can work into the group dynamics. I glance in Julie's direction. Her brown hair flops forward as she scribbles feverishly in the draft of the VOM, taking "orders" from her five group members. They are hammering out nuances of rigging requirements. She is surrounded by two hefty port captains (one of them Dave), one young engineer (Todd), a manager of fleet operations (Will), and Jack, the compliance guy. My team is at a temporary standstill, having completed our content review of the maintenance and repair section. It seems best to check in with Julie to make sure that we are following the writing process. We are trying to accomplish so many things in our two groups that I am already worried about how to integrate all the information back at the office.

I get up from my table, the five "writers" in my group anxious for a break. I have driven the smokers into nicotine withdrawal! I walk across the room to Julie's very vocal table. As I approach, Todd is intently telling Julie something. "No. Pank,

Julie." Julie looks closely for any reference to *pank* in the rigging text. As she examines the book, Todd again says, "No. Pank. I said pank."

With an exhausted sigh, Julie gives up. "What are you talking about, Todd?"

With an understanding look and a smirk, Todd picks up the pink highlighter and gives it to Julie, at the same time taking away her yellow one. "Pank, you should be marking that in pank."

Hearty laughter breaks out at the table as we realize that Todd is speaking in the foreign language of the South. I can tell from Julie's soft expression that she appreciates his input. We had decided early in the meeting to mark mandatory work practices with a pink highlighter and recommended ones in yellow. This distinction is very significant to a worker's safety. Unquestionably, Todd is on his toes, invested in this task.

"How are you doing over here, Julie?" I ask. Julie looks up with tired but still intense brown eyes.

"Don't even tell me that you are done over there?" Julie says with a weary smile.

"No, we were just at a good stopping point. I think that I've pushed Danny over the edge! We finished maintenance and repair, groceries and supplies, reports and records, and have started critical response." I look at her text, trying to get a sense of her process. When she sees me looking over her shoulder, she comments, "I know this looks messy, but I think that I've made good notes about all their decisions. I'm using green to denote moving text from here to other parts of our new library structure. Blue means delete the text entirely; they've determined that it is incorrect or outdated. Does this look like what you are doing?"

"Yeah, you're getting into some real detail here. I think that is because you are working on the operating procedures sections. My team's sections just haven't involved as many actual work practices. Ours are more related to global policy, with informational text. That's why we're moving faster, I think. I was worried that you and I weren't editing and rewriting in the same way. But I don't think that is the case." I look up from Julie to find that her five team members are listening intently to me. Once again, it strikes me how involved they are in the process. They are listening to every word that passes between Julie and me, making sure that we are on the same path.

Julie questions me further on my progress. "I don't understand what you mean when you say that your sections are different than ours. I'm worried that we aren't engaging with the content at the same degree of specificity."

As I think about this, I look at my project lead, Dave. He immediately jumps in. "I see what Carol is saying. The type of stuff we're dealin' with here is very detailed because it's related to the work, to the operating procedures. Take this for example: 'Bridle slings, davit hoist, hoist block, and cable must all be in good condition and inspected prior to use. When the yawl is in stowage or on its cradle, use hold-down straps to hold the yawl in place.' The kind of information that their group is reviewing consists of more general statements, like those that involve company policy on keeping records current or carrying certain cargoes." Julie's puzzled expression tells me that she is still uncertain about the distinction that Dave is making. But, true to the team process, she takes what he said to heart and seems satisfied that we know the VOM better than she does.

My team comes back from break, each one looking at their drafts and getting on the same page. Danny, a captain and a trainer, asks Gary what he thinks about moving all equipment-related information to one place, the equipment book of the library. After staring for a moment at the ceiling tiles, Gary says, "I think it will work, but only if we can determine what we mean by equipment." With his engineering straight-talk, he continues, "To me, equipment basically is something that doesn't breathe. A blue pipe is a water pipe. That doesn't breathe. A deckhand that is supposed to paint the pipe on a regular basis, well, that breathes and probably belongs in the procedures book instead. But I'm not sure that it can always be that clean cut."

As I listen to Gary talk, I again admire his clarity. It is clear to me how he has come full circle in this process. In the first team meeting, he was resistant and grumpy about his role. Now he is supportive of the effort but very clear about possible shortcomings. I like his honesty; you always know where he stands.

"So, how did you think the meeting went?" Using my blue Oldsmobile as an office, Julie and I use the five-hour journey home from Paducah as an opportunity to debrief about our work. Without hesitation, Julie says, "I think it went great. Exhausting but fun. I think that you and Dave picked an excellent time to bring me in for support. It was the right use of my writing teacher/facilitation skills, and it freed you up to be a writer. This meeting is also great for us as a company. It gave you an opportunity to shadow me on working a team through the writing process. This is great timing because you'll need to be working the teams once we start Central Railway's safety rule book next year. Before these meetings, you were growing as a teacher and writer. But now, you've really put that learning into action."

I laugh. "Well, I hate to admit it, but this meeting was fun for me. I was too nervous to enjoy the first team meeting, but this one was different. I could really see the process as creative. I found all the negotiating, problem solving, and communicating to be inventive and dynamic. It seems to bring people's involvement full-circle. I just really enjoy the guys. They are so genuine, so earthy."

I ask Julie, "Where do you want to eat? There aren't many options off this exit." Nothing sounds good to me. She offers, "How about Steak-N-Shake?"

I can see the red sign glow in the mist. Only a few more hours to home, then on to weeks of work at the keyboard. As I brake for the exit, I wonder, with a heavy sigh, if the mountain of vessel operations materials will tumble off the backseat. We had gathered a boatload of information!

Part II

The Process

A professional is someone with expert knowledge, someone not simply with know-what but also know-how. The following nine narratives provide a wealth of insights into the ways in which seasoned professional technical communicators perform their jobs. Some of these narratives trace the days and weeks of complex documentation projects. You will hear writers conversing with SMEs, negotiating with managers, and sometimes talking to themselves about how to solve not only challenging technical problems but also problems imposed by tight deadlines, managers who want to micro-manage, and sometimes writers' own mistaken preconceptions about how to do their jobs. Several stories make simple points, exemplifying situations in which any technical writer may find herself and offering solutions to problems that the authors learned the hard way.

In "Three Months, Three Pages," Rahel Anne Bailie describes the innumerable obstacles a writer can encounter in a large project: computers that freeze up, software failures, anxious managers, and SMEs who have their own problems to worry about. Bailie's story is particularly effective in showing that the ability to create and transform her professional and personal image can be a vital interpersonal skill. Part of this skill, for Bailie, involves playing with gender roles.

Geoff Hart reflects upon the power of the physical environment to shape workplace culture and social interactions. A misunderstanding with a programmer helps him realize that electronic communication systems don't always provide the best means of interaction with others.

Technical communicators need to understand not only complex technologies but also effective networking and teamwork. Mark H. Bloom's story, "Try and Try Again: The Story of a Software Project," portrays in considerable detail—but always in terms of the intellectual challenges, the emotional highs and lows, and the needs of other people—how a development team works together to meet the deadline for a web-based product release. Bloom's story, above all else, emphasizes the importance of adhering to professional standards, continually learning better methods and new concepts, and keeping audience needs foremost.

In "Tech Writing and the Art of Laziness," Tim Casady demonstrates that sometimes even vice can be a virtue when it involves empathizing with users. Casady's story shows how a competent technical writer maintains the user's perspective in nearly everything he does and builds that perspective into his documents. He also has some useful tips for efficient document development.

Julie S. Hile faces eighty recalcitrant railroad employees and managers in "I've Been Working on the Railroad: Re-Vision at BNP Railway." In her story about writing a new railroad safety manual, we see the many skills of a consultant who not only has to quickly win over the members of a long-established professional culture but also actually transforms some fundamental values of that culture. Hile's story shows an approach to collaborative writing, being pioneered in the field of industrial safety training, that goes well beyond current views of collaboration, authorship, and professional responsibility.

In Shawn S. Staley's "Daze at the Round Table," the values of the writer, the programmers, and management appear to be irreconcilable. Like Hile's story, this one raises questions about how far a writer can be expected to go in upholding her professional standards. This story, along with Jong's "Samurai Review" and some others in this collection, also raises questions about typical situations in which technical communicators are not understood or respected and how such situations might be improved.

In "Diary of a Tech Writer," Elna R. Tymes tells about a project on which she served as a contractor who, if not entirely understood, was well respected. Tymes mentions a variety of skills needed by technical communicators who work on large, complex projects, especially freelancers who are hired for their special expertise. Some of Tymes's skills, like baking cookies, are not expected to appear on a résumé.

Steven Jong tells an emotionally tense story of SMEs and a technical writer who have conflicting ideas about documentation. "Samurai Review" dramatically portrays a fairly common step in major document development projects—the draft review. The issues that dominate this story include professional status, relationships with SMEs, the needs of readers and the importance of audience analysis, the influence and the real nature of deadlines, the importance of interpersonal skills, and project management. Although most readers will be disturbed by this story, we hope that it will provoke useful and positive discussions.

In the final story in Part II, "Thumbnails," Rochelle Gidonian describes how she mediates between a technical writer who is opposed, on principle, to including screen shots in Help files and a programmer who believes that screen shots should be included. Only by talking with the programmer and seeing the way that Help files appear on her monitor is Gidonian able to come up with a compromise that meets the needs of users with differently configured desktops. We believe that this story emphasizes one of the most important principles of effective technical communication—understand the user.

Rahel Anne Bailie

Three Months, Three Pages

Rahel Anne Bailie is a technical writer who has produced legal advocacy, computer hardware, and telecommunications material. Rahel transforms her daytime identity as a mousy word nerd to that of an after-hours purveyor of an eclectic assortment of interests, activities, and identities, including amateur musician, ace Scrabble™ player, and grandmother extraordinaire. She lives and works in Vancouver, British Columbia.

It's the third month on the job, and I'm standing in front of a whiteboard. An engineer, his too-young-to-shave jaw clenched in concentration, draws clouds, boxes, telephones, and squiggles ending in hyphenated number sequences. Peter asks if I understand the frame relay network he's just drawn. I rapidly copy the hieroglyphics from the board into my book and nod at the appropriate times but can't answer his pop quiz questions at the end of his demonstration. I haven't quite grasped the concept. I tell him I need to go away and mull over my notes; that's the way I absorb information. He flips the dry erase marker in the air and says cheerfully, "Sure thing. Just ask me if you have any questions. I'll be around."

Telephony. I'm tempted to pronounce it tell-a-phony, like the impostor I am, as I nod, having barely comprehended what this young man, probably younger than my son, has just explained in his rapid-fire style. The basics of networking course I took now seems woefully inadequate preparation for the task ahead. I repeat the mantra of my manager, ensconced in our offices several thousand miles away: they supply the technical content; we just make it usable. We can't write what they don't give us, but what am I being given?

My job is to document a bread-box-size, complex, expensive product used by the telephone industry. The product documentation I inherited is full of errors, drafted by a writer several thousand miles away who relied on hastily written specifications and a photograph of a prototype unit. There are bound to be mistakes in a manual cobbled together that way.... Configuring the unit by using the manual is... impossible.

Back at my desk, I drop my book on the desktop. The hard cover clunks; the gold-stamped company name winks mockingly at me. All engineers are issued such a book in which they are required to enter their activities, observations, and technical decisions. I get one by virtue of proximity. My writing is not the precision script of my colleagues. I've learned to write down the middle third, leaving room for the comments and clarifications the other engineers will offer—like the Talmud, but color-coded: red additions for first-level notes, green for clarifications to the red additions, purple for management overrides.

My friends' eyes glaze over when I describe my job. "Do you realize how many connections have to be made for a single long-distance telephone call to happen?" I ask with earnest awe. "It's absolutely amazing."

"Yeah, amazing," they reply. "Let's go for that coffee."

It's the third month on the job, and I haven't written a single page. Don't get me wrong. I've been busy with everything else but writing. The information systems guys—we affectionately call them the IS boys, as in "we is cool, man"—disappear into the bowels of the telephone room when they see me approach with a user guide in hand. I follow, determined.

"I got the blue screen of death . . . again," I say.

And they grumble, "With your machine, you'd think we'd have fewer problems." It seems that size counts, but tech writers tread carefully when their computers are faster, their monitors bigger, the whole system, frankly, better than the ones handed out to the techies.

Of course, because my software package is nonstandard, the IS boys don't support it. Out of Memory Print Job Cancelled error? "But you've got 128MB of RAM. I'll have to call tech support, but I won't have time until tomorrow. Do you have a *listserv* you can tap into?"

My job is to document a bread-box-size, complex, expensive product used by the telephone industry. The product documentation I inherited is full of errors, drafted by a writer several thousand miles away who relied on hastily written specifications and a photograph of a prototype unit. There are bound to be mistakes in a manual cobbled together that way. The engineers, relieved to hand the manual over to a tech writer and get back to their own work, set me up with a unit. Configuring the unit by using the manual is . . . impossible.

The engineer assigned to answer technical questions scratches his head as he punches at the keyboard. "Oh, you've got the wrong software load," he says, relieved that he's not losing his touch. He asks around until someone gives him the definitive answer about which is the latest release of the code.

I try using the How to Upgrade chapter to install the load myself but fail miserably. I have him walk through the steps slowly as I note the missing steps in the margins of the draft. That chapter will double in length before we're through. The engineer muses about how many assumptions have been left out.

"No one outside the R & D department will know to press 1 instead of F1," I observe, "or that you have to enter this sequence again before entering the channel number, or . . . "

Dave puts up his hand. "Okay, okay, I got it now. Boy, that's some picky stuff, this tech writing. Glad you're doing it and not me."

It's been only a week since someone put a unit on my desk, and I already have an overwhelming urge to open it up and fill it with flowers. Instead, I go to the gym during lunch and work out.

The IS boys have solved the blue screen of death problem, but now it seems they've ordered the wrong software package, not realizing the difference between FrameMaker and FrameMaker+SGML. The purchasing agent didn't seem to realize the urgency of ordering the new software—she tells me the requests from the R & D engineers get priority, and she's swamped right now—so the writer several thousand miles away will make have to make the changes in the files. I document the errors and suggest wording, then e-mail the week's discoveries to her.

Tommy comes to my desk with the section of the manual he's been given to review. I hear he's a voice compression genius. "This table is wrong, you know. I shorted out three modules using these directions."

I blanch. "So why didn't you say something before now?"

"Didn't occur to me to tell you," he shrugs. "I didn't know this is what you were writing."

My look gives away my amazement that my colleagues, so skilled in their area of expertise, don't make the connection between what they develop and the documentation. I ask what the correct information should be.

"Isn't that your job?"

"But I'm not an engineer," I remind him. "It's your job to give me the technical content. It's my job to make it usable. And why, if you install these modules in chapter 2, do we have them documented in chapter 4?"

"I don't know. You mean you need to put stuff in the order people do it," Tommy laughs. "I thought we'd put all the tables together so it would be easier for me to see what I'm doing."

I roll my eyes. "No, no. Unclear on the concept, Tommy. You provide the puzzle pieces. I figure out how the puzzle fits together."

Tommy cocks his head and, working the logic out in his head says, "Then what you do is more like rewriting what we write?"

I concede rather than argue. I'm already behind schedule. Instead, I go back to poring over the hundreds of pages of densely packed facts, ferreting out inconsistencies and making suggestions about structure. I don't know if we'll have time to improve on the cosmetics.

It's the third month on the job, and I've held two technical review meetings, bought four dozen doughnuts, and made one funny birthday card. The latter two items are bribes—incentives, if you're practicing corporate speak—for the engineers to cooperate in the first item, the tech review meetings. I'd be lost without them. The project manager disagrees, but I get my director's permission to insist that all the engineers who worked on the manual get locked into a boardroom for two consecutive mornings. Just them, me, and a dozen doughnuts. We go through the manual page by crackling dry page while I have them swear the content on that page is correct. To emphasize my point, I write *okay* in the top corner of each page in thick red marker and circle it. If a customer can't make something work at some remote site because of incorrect instructions in the manual, I certainly don't want to take the heat.

Bringing food to engineers is a time-honored tradition. It makes them detour by your desk for a midafternoon sweet-tooth fix and brings out the eloquence of even the most recalcitrant engineers. I celebrate the completion of a manual by picking up a large box of miniature hot buns, half sweet, half meat-filled, from the Chinese bakery on the way to work. When I stop at the coffee station for my morning dose of caffeine, heads pop up over cubicle walls. I write, "Thanks, guys, couldn't have done it without you" on the back of a business card and tape it to the inside of the lid of the box. The buns will be gone within the hour.

It's the third month on the job, and I've taken a chance. The human resources staff are trying hard to be the cheerleaders of this cynical bunch of engineers. The three HR gals are dressed up in costume: a witch from the Wizard of Oz, Cinderella, and one of the characters from Star Trek. Scully, I think. No, wrong show. I should know which character, but I'm too embarrassed to ask; admitting my ignorance would be a considerable setback in my efforts to be part of the gang.

There are supposed to be prizes for best costume, but none of the guys have dressed up; though when asked, they will tell you their imaginary costumes. "Can't you tell? I'm Arnold Schwarzenegger in *Terminator 2.*"

I decide to devise an impromptu costume. Retrieving only slightly smelly white socks from my gym bag, I put on the socks, roll up my jean legs, and hitch them up as far as they can go. I tuck in my polo shirt, embroidered with the company name, and tape a pocket protector across my chest. Dave suggests I put some Scotch tape across the bridge of my glasses. We emerge from our corner and claim a spot at the intersection of two aisles between the rows of cubicles. We pretend to talk about work while watching the reactions of passersby. Wayne saunters by, looks at me, and dissolves into hearty laughter. He keeps going, back to his cubicle, but turns every few feet and starts laughing all over again.

Peter walks by, head down, absorbed in his assembly language tools manual. He glances up and says, "Broke your glasses, huh?" and returns to his book.

Dave gapes at Peter, then turns to me and chuckles. Other engineers begin to come by. One gives me a pencil for behind my ear. Another tucks a heavy-duty calculator into my belt.

Peter comes by again, looking sheepish. "I heard you dressed up like an engineer. Sorry I didn't notice before. I just thought you broke your glasses."

It's the third month on the job, and I've got some buddies now. It turns out that I worked with one of the engineers eight years and two cities ago. Back then, we worked for a stodgy multinational firm; now, we work for this vibrant start-up company. He, my cube buddy Darby, and I go out for sushi on Tuesdays. They show me how to fold origami chopstick rests from the paper sleeves covering the chopsticks; my role is to interpret what their girlfriends say into what it means to them. They think I'm a bit of an oddity, but I think it increases my cachet. I've carefully let out my line, revealing my personal self bit by bit. It's hard enough being one of the few nonengineering staff. I don't want to scare them off by admitting to all my differences. First, I let on that I was quite a bit older than them, then told them the car seats in my minivan are not really for my children but my grandchildren. Once they got comfortable with that idea, I revealed the next-safest detail.

I imagine them talking together out of my earshot about what a strange duck I am. Maybe I'll dye a patch of my hair blue next week, just to keep them talking.

I'll remind them, tongue in cheek, that blue is our corporate color. Blue hair will blend in with the blue jeans and blue polo shirts I've taken to wearing to work. We all dress this way, except in the summer when the guys with toned legs from rock-climbing come to work in shorts, T-shirt, and Teva sandals. After the first week, I retired my wardrobe of skirts and suit jackets to the spare closet and replaced it with a half-dozen company polo shirts and two pairs of blue jeans. I have a pair of chinos I wear with dress shoes (i.e., not running shoes) on days we have meetings, just like the guys. My boss, a motherhood-and-apple-pie kind of guy, has never quite figured out what to make of me; he probably thinks all tech writers come this way, but enjoys my company in an odd sort of way.

It's the third month on the job, and I've made the grade, though a discovery I've made has caused the schedule to slip. I've gone through the manual, methodically documenting a new feature for the tech writer a few thousand miles away to incorporate into the manual. The engineer who designed this feature has been pleased with my capacity to understand the concept, so I have not been too intimidated to ask him what seem to be dumb questions. When he told me there is no such thing as a dumb question, only a dumb answer, I was encouraged. Now I explain my dilemma in documenting the new feature.

"When I explain here," I thrust a finger at a table midway down the screen, "that the user puts in this information, it works in line one and two, but if they put the information the way I show it in line four, then I can't figure out how they'll stop going in a circle."

The engineer, staring at the screen, looks lost in thought. "I see," he says, and bumbles away back to his desk.

I stare after him, wondering what his departure means. Did I ask something so incredibly idiotic that he couldn't be bothered answering me? Or has he gone away to think about what I said? Should I go after him? Should I ignore the entire conversation and leave the documentation as is? I grab my coffee cup and head to the kitchen. When I get back to my cubicle, he is heading toward my desk.

"It's good you found this problem," he pronounced, rather formally, in his heavy eastern European accent. "I did not realize when I designed this what could happen here. It will take me some time to fix the problem. Maybe you want to work on some other section of the book until I find a solution. You see, this problem affects some other features, too, so the next four chapters you have in your book will also probably change."

At lunch on Tuesday, over tuna maki and dynamite rolls, my buddies tease me about finding the bug in the software. I smile but am more worried about the implications for the publishing schedule. How can I get the manual to the printer on time if the rest of the schedule slips?

It's the third month on the job, and we're being bought out. Management makes the announcement on a Wednesday morning. The front of the boardroom is filled with executives from our future company, and our managers stand off to one side. A bedraggled bunch of engineers listen to the news. They have been at work until close to midnight, trying to finish up the last of the design verification tests, then came in again at eight o'clock to do it all over again.

The younger guys, many of them in their first job out of school, are in shock, while the more experienced and cynical of us begin mentally calculating what the implications will be for our careers, our colleagues, and our paychecks.

"Just when you got your computer up and running, too," says Darby.

"If they hadn't canceled the project, it would have meant a simple change to the FrameMaker header variable. Okay, so I write about the product for this company instead of that one," I counter. "Or I write about a new product."

My boss puts his arm around my shoulder, and we both laugh at the irony of the situation. "Now that the IS boys got me set up so I can actually get some writing done, I produced a whole three pages this week." I shrug. "I guess those three pages will be the only ones I ever write for you."

Geoff Hart

Conquering the Cubicle Syndrome

Geoff Hart says, "I've been writing and telling stories (occasionally getting into trouble thereby) since I was six, though I was twenty-five before I realized I could make a living at it. The lack of a career path in forest biology helped me see the trees amid that particular forest. Fortunately, IBM was intrigued by my apparent lack of qualifications, called me up to wonder why I'd wasted their time with my résumé, and hired me that afternoon; either I'm very persuasive, or they were very desperate. Soon after donning the salaryman's blue uniform, I ducked a massive re-engineering and found a home with the Canadian Forest Service, where I toiled happily for six years. Keeping an ear to the ground led to dusty ears but helped me bail out just before a rightsizing and return to my home town of Montreal. I've been working there ever since, doing technology transfer for the Forest Engineering Research Institute of Canada.

"I've worked as a technical communicator since 1987, doing editing, technical writing, French translation, creation of online help, website design, video scripting, slide presentations, and information design. I also do Windows, Macintosh, and (rarely) UNIX. I've considered adding a hat rack to get all those hats off the desk, but fear that They might add more hats.

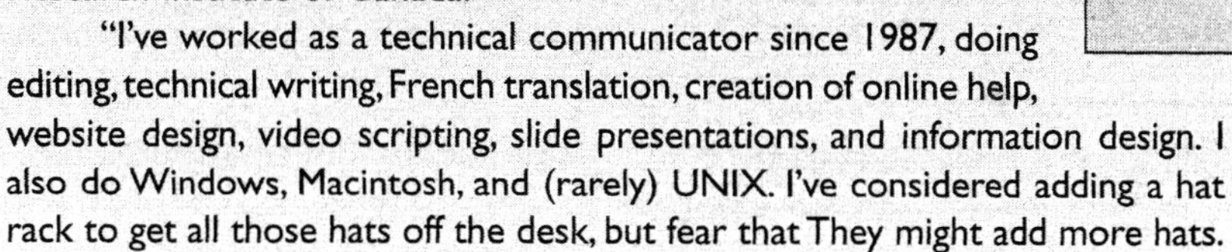

"My two kids (Matthew the Entomologist and Hurricane Alison) keep me busy at the tender age of thirty-eight, but leave me time for weekend hockey with a depressingly young crowd, for various sweaty daily exertions intended to help me survive hockey, and occasional freelance jobs for a Japanese client. In what little time remains, I write for the Society for Technical Communication's newsletter (*Intercom*), the 'User's Advocate' column for *Computerworld Canada,* and several novels . . . when I'm not occupying an altogether unconscionable amount of time and bandwidth on the copyediting-l and techwr-l discussion groups.

I work in Quebec, where business tends to be conducted in a strange mixture of English and French that we locals refer to as "Franglais," but you'll find similar pidgin dialects anywhere that English speakers interact with people from different linguistic communities (i.e., engineers, programmers, and scientists).

I'm going to violate one of the cardinal rules of technical writing and alienate you—my audience—right off the bat: I've never worked in a cubicle. But—and here's where I try to regain your sympathy—I *have* worked for long periods of time under similar circumstances, and what I humbly dub "the cubicle syndrome" extends far beyond those domino-style wall panels technical writers love to hate. Still with me? Good.

So what's this cubicle syndrome I referred to? It goes a little bit like this: cubicles aren't really physical walls—they're a state of mind, in effect, the sense that you've been compartmentalized and isolated and pigeonholed. You know all about cubicles if you're a Dilbert fan (and most technical writers are, for reasons you'll discover once you've been in the business for a while). Cubicles are those four-sided, felt-lined livestock pens loved by evil office managers everywhere, but that description hides the Platonic ideal that underlies the structure: cubicles are really all about isolation and becoming part of the building infrastructure, whatever the actual physical environment of your chair and desk.

For instance, my colleagues at my current employer dubbed my first office "the aquarium." I remember the first time I heard that name, as I was just settling in and filling my desk drawers with the usual assortment of pens, Post-its, and paper. Simon, our graphic artist, stuck his head in the door, a big smile on his face.

"So how do you like the aquarium?"

"The aquarium?"

"You'll find out."

The office was about eight feet square, with just enough space on the far side of my pressed-wood desk to fit a chair for my guests, squeezed between the far wall (and I use the term "far" metaphorically, you understand) and a filing cabinet. It had four barren, industrial-blah-colored walls that might have been attractive when they were first painted but had long since lost whatever vigor they'd possessed in their youth. My very own door hinted at the potential for privacy, but it was only a hint, because the room's only window was a floor-to-ceiling glass panel beside the door that looked out onto a high-traffic hallway, letting in a pale shaft of fluttery fluorescent light. The window's one pretension to modesty was a curtain so diaphanous that it would have violated local obscenity laws if anyone had tried to wear it publicly in place of clothing.

In short, my little *pied-à-terre* had all the drawbacks of a cubicle (a small, claustrophobic space with no view to speak of) and none of the advantages (warm, fuzzy walls and neighbors you could peek in on or gossip with over the tops of the partitions). Worse yet, people walked past constantly, and many paused to stare in through the window and see what the new guy was doing. I quickly came to empathize with how the fish in an aquarium feel, in all respects but one: because I wasn't a subject matter expert (the legendary SME, pronounced "smee"), no technical writers ever dropped in unannounced bearing food in an attempt to distract me from my work and thereby buy them time to question me.

We moved to a brand new building within a year, so I wasn't really there long enough to ever grow comfortable with that place. I did eventually hang enough posters and add enough detritus to my desk to make it "home," but it was still a sterile environment at the best of times. So much so, in fact, that I made a point of escaping whenever the opportunity presented itself and paying a visit to my new colleagues, thereby getting a feel for whom I was working with and how best to work with them—and returning the favor so they'd know whom *they* were working with. It's particularly important to build those contacts if you're working as an editor because you're going to be treading on a lot of egos with your revisions, and that treading is a lot more bearable if they know it's being done by someone they like and trust. Those relationships are just as important if you're doing technical writing because people who like you and respect you are far more willing to make time to answer your questions and review your writing.

When we moved to our new building, I traded in the aquarium for a room with a view. My new office is painted a soothing shade of warm gray, with a matching carpet. It's larger too, nearer to twelve feet square, with enough room for bookshelves and three desks. You might think that one desk is sufficient for most people, and you'd be right, but for me, the two extra desks turn out to be no luxury. The first desk holds my computer, an ever-expanding row of software manuals (I'm one of those rare birds who actually reads the damn things), and miscellaneous scraps of online help projects and manuscripts undergoing online editing; I spend an increasing amount of my day there now that we're moving slowly toward an online writing and revision workflow. My second desk, originally intended to provide room for me to work side by side with authors, now holds my in basket plus the overflow from the never-ending stream of paper that flows into my office; I'm living proof that the "paperless office" isn't going to happen in our lifetimes. My third desk is where I do on-paper editing, and though it originally offered a respectably large workspace, what's left after several years is a space about one foot deep by two feet wide in which I can actually work. That workspace is surrounded by piles of neatly organized papers (that's my story, and I'm sticking to it) that surround me within easy arm's reach. If it's true that a tidy desk is symptomatic of a sick mind, then I'm one of the sanest individuals you'll ever meet.

Best of all, I now have a door I can close if I really need some privacy, and a real, wall-to-wall window that looks out onto our backyard. There are young trees in the foreground, planted a few years back when they finished construction of the new building, plus older trees a bit farther off against the horizon and a soothing expanse of grass in between. Sparrows, squirrels, groundhogs, and one neurotic cottontail rabbit live there, and the occasional cat crosses now and then like a lion on the Serengeti. There's even a mob of ravens that occasionally descends on the yard for a conference. I've finally got enough light to sustain a small jungle indoors too, though the only wildlife I've seen so far has been the occasional spider who came inside to escape the Canadian winter. Very nice indeed! Now if only I could open that window . . .

But it turns out that I've inadvertently fallen victim to the sinister effects of the cubicle syndrome. I now had all the advantages of a real office—and a very nice one indeed—but the isolation was still there. I now had neighbors (my communications team colleagues) and that must have been what lulled me into a false sense of security, because those neighbors unfortunately weren't the authors and SMEs whom my life as a writer, editor, and translator depended on. I rediscovered this crucial distinction during the first online help project I undertook.

It was early in the project, and I was still just playing with the software, getting enough of a feel for it that I was beginning to understand how I'd have to organize my manual and online help, and not late enough that I actually had to start writing. Because I'd been parachuted into the project midway through the beta testing, the software was nowhere near ready for prime time, and I was keeping careful watch for bugs as I played. Round about noon, after a few hours of experimentation, I'd used the software to build up a respectable database of information and an equally good understanding of the software. I reluctantly decided it was time to start a new database file and begin documenting how I did that. To verify whether the appropriate error checking was in place, I chose to give my new working file the same name as the file I'd been working on all morning, fully expecting a solicitous "Are you nuts?" from the software. Unfortunately, that was one error check the developers had neglected to include, and the software obligingly overwrote and erased an entire morning's work. I wasn't particularly heartbroken, you understand, since that work had only involved testing the software rather than creating something my job depended on, but I imagine our future clients wouldn't have been so sanguine. So I phoned up the developer responsible for that module. Jean works one floor up and maybe one hundred feet away from me as the ant crawls, assuming the ant is a tourist and doesn't take the direct route.

"Jean, I think I've just discovered a pretty major bug. I just lost three hours of work in the Harvesting module because the software didn't warn me I'd be overwriting an existing file."

Dead silence on the other end of the line. Just as I was beginning to get worried, Jean cleared his throat, and in a somewhat hurt tone, replied, "I don't understand. Why do you call this a bug?"

I experienced one of those wonderful moments, ubiquitous during encounters with SMEs, when you suddenly feel as if you're the first human to encounter an utterly alien race, and you have no idea how to communicate with them. If you work for any length of time in technical communication, you'll encounter such moments often enough; this time, the problem was exacerbated by Jean's being French and my being English, and that adds a certain linguistic and cultural barrier on top of our already very different working contexts. I work in Quebec, where business tends to be conducted in a strange mixture of English and French that we locals refer to as "Franglais," but you'll find similar pidgin dialects anywhere that English speakers interact with people from different linguistic communities (i.e., engineers, programmers, and scientists).

The problem, it turns out, was not a communications failure per se so much as the sinister effects of the cubicle syndrome. You see, I'd been enjoying the privacy of my new office so much that I'd been actually enjoying my stay there and somehow had begun "cocooning." In short, I'd stopped getting out of the office as often as I should have and mingling with the authors and SMEs as I'd done while I was condemned to my old office.

I solved this particular problem the same way I probably should have solved it right from the start: I made the heroic effort required to leave my "cubicle" and walk the one hundred feet necessary to pay Jean a personal visit. That's admittedly an arduous effort, but fortunately, I walk to work every day and play hockey on weekends, so my heart was up to the challenge. (If you choose to do this on your own, please see your doctor to be sure it's safe for you, and ease into the exercise program gradually.) Once I reached his office, Jean and I chatted about various things for a bit. Then I reintroduced the problem, and it turned out that we'd simply tripped over a difference in our respective jargons: to Jean, a bug was a calculation error, not something that caused lost files. We certainly both agreed that the problem needed to be fixed; it was just the terminology that was problematic.

Round about then, long after my brain should have kicked in, I remembered that Jean also ignored written questions on his edited manuscripts and that the only way to get good responses was to drop in and ask those questions in person after warming him up with some human contact unrelated to work. Armed with this new understanding, and the knowledge that Jean was simply one of those people you have to talk to face to face to communicate successfully, I had a much easier time with the rest of the documentation project.

I also took home a much broader lesson, one that I'd already learned but had somehow forgotten along the way: *communication is about contact between two people, not simply an exchange of words.* Some people are perfectly happy to communicate by e-mail, others in print, and others by phone, but many simply prefer to communicate face to face. I've once again made it standard practice to informally drop in on my authors and SMEs at least weekly—more often if we're collaborating on a project—to ensure that the lines of communication remain open and that I don't just drop in when it's time to assault them with a problem. The ongoing relationships really do smooth out the more formal business aspect of our relationships.

And that's the real take-home message: a cubicle *really is* a state of mind, and (you should pardon the wording) thinking outside the box is the only sure cure to the cubicle syndrome.

Mark H. Bloom

Try and Try Again

The Story of a Software Project

Mark H. Bloom is an award-winning technical writer and an acknowledged expert in online help development. Mark currently manages a documentation team for America Online, Inc. An active member of the Society for Technical Communication, he has given presentations at international events and has served as president of his local chapter, an organization with over four hundred members. But it's not all work and no play for Mark. He is an avid baseball fan, a motorcycle enthusiast, and a student of life. Originally from Salem, Massachusetts, he moved to California in 1995, where he and his wife enjoy the climate and the culture. As part of a writing consortium named Adventure Architects, he collaborated on three books published by the gaming industry. He is finally working on his first novel, and he fully expects to have it published.

In the beginning, I believed it was going to be a relatively straightforward project: create an online Help file for novice computer users. The program to be documented was a WYSIWYG, point-and-click, step-by-step tool for creating a homepage. No HTML experience was necessary to use this program since HTML never slips past the user interface. Users simply pick a sample page and then replace its content with their own, piece by piece. What could be easier?

The program, which I'll call HomePage 2.0, was originally intended to replace the older, more limited HomePage 1.5. The HomePage 2.0 project, then, was a simple software upgrade. Unfortunately, the direction of the project would change in midstream, but I'll get to that later.

> *The team considered my ideas as carefully as they did anyone's. It's a rare situation in the technical communication field—a writer being an integral part of a development team—but I have fought throughout my entire career for the opportunity.*

As the technical writer, I was part of the development project team. We would meet every Tuesday afternoon for about an hour to discuss plans, progress, and problems. The team consisted of Eric the project manager, Linda the development lead, Jeanette the user interface specialist, two programmers, three quality assurance (QA) testers, and me.

I joined the project immediately after its alpha release, so the team had already completed some initial coding and design work. I installed the alpha build easily and got to work. My first step was to dive into the product—determine what it was supposed to do, see what it could do, and decide which questions I would have to answer.

I gathered as much existing documentation as I could find, hungrily collecting it from any source and poring over it with relish. There were design diagrams, marketing specs, and feature lists . . . all of them preliminary and indistinct. I needed more.

Meeting the Team Leaders

I decided to meet with the leads on the project team. Only by talking to them all could I judge the scope of the job ahead of me. I knew from experience that no one individual would have all the answers or the best perspective. The QA folks, for example, would know whether the features worked, but not necessarily how much users would need a certain feature. The good news was that everyone was handy . . . and friendly.

I first met with Eric, the project leader. He was in charge, so I wanted to get his expectations of my role and task within the project. We met in his office. He sat comfortably behind his desk, while I sat in a stiff upright chair with a notebook on my lap. Still, he devoted his attention to me. What impressed me most was that while we talked, while our informal meeting took place, he did not answer his phone or look at his computer monitor. His world was on hold while he met with me.

I got right to the point. "What kind of documentation does this project require, just online help?"

"That's correct. Since our customers will download the software, it doesn't make sense to give them an extra file they won't look at anyway. The online help should be able to answer all their questions, don't you think?"

"That's my goal." I was disappointed about not being able to write a good user manual for the product, but Eric took the decision out of my hands. I tried to make the best of it and focus on the Help file. "What about context-sensitive help? Can we link some into the application?"

"What do you mean by context-sensitive help?"

"Well, it's linked to the user interface, with a Help button or a Help icon . . . "

"Oh, isn't that how all online help works?"

"Usually, there's a Help menu or . . . "

"The HomePage program won't even have a menu bar. It's merely a series of screens, each designed to complete a single step. You'll have to deliver all the help from the user interface. You can do that, can't you?"

"Hold on." I was writing furiously into my little notebook, trying to catch up with the conversation. I had to write it down and try to process it at the same time, but I saw where he was going. "Sure, I can do that. What can you tell me about the intended audience?" I asked.

"Do you watch television?"

I played along. "Occasionally."

"Well, we like to think of our audience as the Homer Simpsons of the world. If Homer can figure out how to use HomePage, then we've succeeded. We're talking neo-novices here. Some of our users might want to have a webpage simply because someone has told them they should have one. They might not know anything at all about webpage design, structure, or convention. They might not even know what it is they want."

"Sounds like a challenge."

"It certainly is. We have to present the whole program in easy-to-read segments. Marketing wants no cryptic error messages, no leaps of faith, no false assumptions. Everything—including the online help—has to be developed with Homer Simpson in mind."

I was writing furiously. "Okay, I can do that," I said. "After I talk with everyone else, I'll figure out what's important and present it so that even Homer Simpson can understand it. That's my strength."

"Excellent. I'm looking forward to seeing your work."

"Is there any chance I can talk to the marketing representatives on the project? It would help to get their input . . . "

"Ah, unfortunately, that's not possible. They are located in Houston, and they have many other concerns besides the HomePage project. I suggest you give me any questions you have for them, and if I don't know the answer, I'll find out for you. Will that work?"

I had no choice but to consent. Eric wanted to be the liaison between the development team and the rest of the company. I could understand his desire to control the project, even if I didn't agree with it. The decision cut off some of my resources, even if it gave me a single contact for any problems or questions that might arise. We talked a bit longer about the project's time line, and then I left him to return his phone calls.

I went directly from his office to the QA area. Since I've worked on other projects in the company, I knew many of the QA testers fairly well. They are a rowdy bunch, the loudest part of the building, always the last to leave a company function. We get along superbly.

I found Robin, the QA lead, amid the maze of cubicles. Her space had just enough room for the built-in desk and an extra chair. We sat and chatted awhile before getting down to business. Robin was as excited about having a technical writer on the project as I was about being there. Although we had never worked together on a project, she had heard about my work on another project. Reputations travel quickly within one building.

Since I do some testing as part of my job—using and researching the application I have to write about—Robin and I agreed that I should notify her if I found any bugs. I had access to the bug database, so I could check whether my bug was new or not. She wanted to double-check it herself before writing up the bug report. I quickly agreed; it meant less work for me.

Robin agreed to put my name on her e-mailing list, so I could keep up with the bug fixes, too. Finally, I convinced her that I should edit the Readme file after she put it together. My feeling is that I should at least review anything that goes out to our customers.

Next, I tracked down Linda, the development lead. I found her in the office of one of the developers, talking about a particular code reuse strategy with her whole team. I waited while she finished her discussion.

"If we take a modular approach," she was saying, "we can reuse whole sections of code. If we end up having to make a change, we only have to make it in one place for it to affect everything. Now, the first thing to do is to identify the code fragments that we can reuse most frequently and start working on those first."

Another developer then said, "Shouldn't we have done that from the start?"

"Well, sure, but it's not too late. The UI is going to change anyway, so there's plenty of time to swap code in and out . . . "

My ears perked up when I heard her mention the UI—the user interface. Code reuse won't affect me at all, since it won't affect the way the application works, but the UI is central to my task; it's what I'll be documenting. From an impromptu meeting to which I was not even invited, I discovered a gold nugget of information. I'd have to dig up more about this later.

When Linda finished and the discussion broke up, I asked her if she had a few minutes. She did.

I followed her back to her office, and we sat around a small round table. While she took a long drink from a bottle of spring water, I began. "I am going to write a Help system for HomePage, and I need to be able to tie the help into the application's user interface. I had a look at the original specifications, and I examined the original UI design, but did I hear you say it was going to change?"

"Yes and no," she said, then paused. "Have you talked to Jeanette yet?"

"No. She's next on my list."

"Well then, I'll let her explain it to you—she's more plugged into the UI than I am."

"Okay." I directed the conversation back to the help. "Since there won't be a menu bar—that won't change, will it?"

"Not as far as I know."

"Then I assume there will be a Help button on every screen. I'm going to need help from a programmer to get the context-sensitive help connected to these buttons. Whom should I contact as a resource?"

She thought for a moment. "Try Chuck. He'll be working on the front end, tying the UI into the program."

"Thanks." I made a note, then asked, "How often do you think you'll be releasing builds?" I wanted to get an idea of how quickly the application might change.

"Initially, we plan to create a build every other week, but after the application stabilizes, we'll be putting out builds at least once a week, maybe more frequently than that." She paused and took another drink from her bottled water. "As you know, the functional requirements are not quite set, so we're still fighting with marketing about what the product has to do."

"I understand, believe me." Marketing was notorious for provoking "feature creep," when new features are added to a product after its design phase. "Are you planning to send e-mail when your team releases a new build?"

"Oh, that's a good idea, isn't it? Yes, I'll make sure Ned does that. He's our build coordinator."

"What's a build coordinator?" I felt silly asking, but a technical writer has to be able to communicate. This includes being able to listen and ask questions, no matter how silly, until you know everything you need to know.

After a quick gulp of spring water, Linda replied, "Ned makes sure that he has all the updated files from all the developers working on the project. Then he copies them onto a clean computer and runs the build." She noticed my confusion and explained further. "He actually builds the application from all the source files. When the build is complete, he does some preliminary testing to make sure the application can run. After he puts the application up on the network, I'll have him send e-mail to everyone in the project, including you and QA."

"Great. Thanks, Linda. That's all I need for now. If I need any more information, whom should I ask?"

"For now, come to me. I know what everyone's working on, so I can direct you to the source if I don't have the answers."

"Okay. Thanks again." I left Linda to her bottled water and went to find Jeanette.

I had never worked with a user interface specialist before and was a bit apprehensive about it. Would she usurp my position as a user advocate? Would she discount my ideas for the product's interface? This was one of the reasons I had left meeting Jeanette for last.

Jeanette was in her office, surrounded by papers. Every square inch of horizontal space, except the floor, was littered with diagrams, notes, and printed e-mail messages. She was intent on her computer monitor, which displayed some kind of drawing program. She didn't hear me enter her office, so I knocked lightly and asked, "Anyone home?"

She spun around in her chair, saw who was invading her privacy, and then smiled. "Mark," she said, whisking a bit of stray hair from her eyes, "glad to see you. Come in and have a seat." She popped out of her chair to clear the debris off another chair. As I settled in, she returned to her chair. "So let me guess," she continued. "You're here to take the test."

"The test?" I asked.

"You're not here to take my usability test? Didn't you get my e-mail? I think I sent it to *everyone.*"

"Sorry to disappoint you. I stopped by to talk about the HomePage project. I'm going to be writing the online help, and I thought you might have a few minutes to . . . "

"Oh yes, that's right. This should be interesting; I've never worked with a technical writer before."

"What a coincidence. I've never worked with a user interface expert before. What is it that you do?"

"Let me show you." With that, she gathered one of the paper piles and gently placed it on the edge of the table, right on top of whatever was already there. Then she flipped the top sheet over, so that it faced me. The page held a rough drawing of the welcome screen of the HomePage application.

I studied it a moment and realized this screen was slightly different from the one I've seen. I pointed out the differences, but Jeanette waved my comments away. "There will be more changes to come," she said. "The functionality should remain the same, but the graphics and the design will change week to week."

"Oh," I said. "Does that mean the buttons on each screen won't change?"

"Yes and no. The look of the buttons might change and the placement of the buttons might change, but the same buttons should still appear on the same screen." She paused and pointed to the page that lay before me. "But what do you think of this screen? Would you be able to figure out what to do next? Is everything clear?"

I examined the drawing again, pretending it was program on my computer. "Yes, it looks fairly straightforward. I'd click here," I said, pointing to the Next button.

"Excellent," said Jeanette. "And then you would see this screen . . . " She dug into the pile and pulled out another drawing. "What would you do now?"

This continued for some time. I examined each screen she showed me, made comments, and gradually worked my way through the application. Although I knew what I was supposed to do, the drawings showed me a different face to the application, so I had to study each new screen. Jeanette made some notes while we worked, but did not comment on my progress or choices.

When we finished, she said, "That is what I do. I work with the flow of the interface to make it intuitive and simple. Everything has to be right—the choice of words, the position of the buttons, and the way the functionality is introduced. And thanks for taking my test."

I realized she'd pulled me in without my consent, but I knew the time was well spent. All I could do was smile back at her. "You're welcome. Any time."

"Be careful what you offer. I might just take you up on that. Any time." She smiled again. "So what do you do as the technical writer?"

I told her about my role in the project. I wanted to create online help that would solve problems and answer questions. To do that, however, I needed to know what the user interface would look like. I needed to know how the user would access the functionality that the application provided. I needed to be able to anticipate what questions the user might ask.

Jeanette was sympathetic to my needs, but told me that despite all our efforts, the UI would change more than once as the project developed. "It's not just my doing," she said. "Marketing is trying to get a certain 'look and feel'—to use the industry buzzwords—and I'm not in a position to deny them. What I can do is keep you in the loop when I see changes are coming."

"That's about all I can ask for at this point. Thanks, and I'm looking forward to working with you."

"Same here, Mark. Thanks for stopping by." She whirled around in her chair, back to her monitor.

Getting to Work

I returned to my office and reviewed my jumbled notes. Then I got down to work. I wrote down as many major topics as I could. I knew I could add minor topics later, but I needed a solid base from which to start. I first listed the procedures I would need to document, such as *Designing your page* and *Adding a picture.* I included every major task in the process, as well as some of the supplemental ones (like *Formatting your text*). Next, I listed the definitions for terms such as *homepage* and *template.* I included every term and concept that I thought a beginner might not know. The list quickly became extensive.

Once I had two fairly complete lists, I had to implement an organizational structure for the help system. Using pen and paper—because I didn't have the software—I drew a flow chart diagram that began to link one major topic to another. Because of the way the application worked, screen by screen, I let the help follow suit. Although users might well perform the tasks in a different sequence, each task was essentially self-contained. Plotting which topics to link was a relatively straightforward process.

Now that I had a list of topics and knew how to string them together, I had to make some other decisions before I started to write. Delivery and formatting were as important as content.

I decided to create a Windows 3.1 Help file. Even though the program would run only on Windows 95 computers initially, the team would be developing a Windows 3.1 version next, at least according to the project schedule. A Windows 3.1 Help file would work on both platforms, so that would be less work for me and have no impact on the users—a real no-brainer of a decision.

Since the program offered online help only by way of Help buttons, I had to figure out which topic would appear for the Help button on each screen. The programmers and I would then link each Help button to the help topic specific to that particular screen. It was ingenious. Point-and-click simplicity.

The simplicity had its implications. For example, the Help file's table of contents (TOC) suddenly became much less important. The only time users would see the TOC would be if they specifically clicked the Contents button in the Help window. If users saw the information they needed right away, chances are they would not need to go to the TOC. This struck me as a good thing.

I decided to link each program screen to a main help topic that would display or link to all pertinent information about that screen. I knew I would invariably have to include some topic links, which could confuse a novice user, but the basic structure was simple—and sound. This structure would allow the user to "drill down" for more and more detailed information, depending on his or her needs. It would take hard work and planning to do correctly, but it was definitely the best approach.

An idea was born!

With the structure in place, I began writing help topic content. I created a simple style for procedures, bulleted lists, notes, and warnings. I used graphical elements to draw attention to important information. I reproduced the buttons and other graphics to tie the instructions back to the user interface. I wrote, I edited, I formatted. (Say *that* in Latin!)

Working within the Project Team

Meanwhile, I participated in the weekly project meetings. In these meetings, I was just "one of the team members." As everyone gave their status updates, I noted any user interface changes or any other developments that could affect my documentation. When the need arose, I voiced my opinion. The technical writer is the end-user's best representative on the development team. If a feature appeared to be too complex, I said so. If I had an idea that made sense, I presented it. The other members of the team always listened thoughtfully to my ideas . . . before insulting me and my heritage.

No, that's not true. The team considered my ideas as carefully as they did anyone's. It's a rare situation in the technical communication field—a writer being an integral part of a development team—but I have fought throughout my entire career for the opportunity. It just makes sense. Why ignore the user's voice in the product's design? The technical communicator is in the best position to know how the user might feel about a specific application feature, an interface device, or even the wording on a dialog box. I made it a point to voice my concerns and recommendations.

Not everyone can carry this responsibility. First, you need good ideas, which only experience can provide. You need to be self-assured without being arrogant. You need to be vocal without being aggressive. I rely on my sense of humor. I'm not a class clown, but I have seen how humor eases the tension and dissipates the boredom in a meeting. Humor also helps me befriend the other team members. They respect my work only through the contributions I make to the project, but they like me because of everything I add to a project team. If I add nothing to the project, then I'm an unnecessary distraction, but if I add value to the project, then I add value to the team.

During one particular Tuesday afternoon meeting, we were all trying to figure out what should happen if the program terminated prematurely. The discussion slowly degenerated into a heated debate. . . .

Eric: We have to allow the users to do Save or Quit any time. That's a requirement. Are you telling me you can't do it?

Linda: It's a disruptive operation. Look, if the program stops, there is no simple way to mark the file and return to the same page later. The page is essentially blank until the user populates the fields with his own content. Even if the user saves the page before quitting, it still would have only the content he added to it. An unfinished page would look awful.

Jeanette: Besides, how would the user know how to Restart a page? The program was designed to take the user from beginning to end, not from the middle to the end.

Eric: But what would happen if someone quits in the middle or turns off the computer? There has to be some kind of contingency plan. You can't just say, "Sorry, start again."

Mark: How about, "Thank you, sir, may I have another?" Seriously, is there any way to detect a page that was saved halfway?

Linda: No. All saved pages appear the same to the program.

Mark: Then how about somehow directing the user to *edit* a page he has not finished?

Jeanette: That won't work either. To be effective, the program has to step the user through the page to add each object. If you ask the user to edit a page that he didn't finish, he won't know where to start or what to add. Editing a page is difficult enough with a *finished* page.

Eric: Well, we have to hash this out somehow. It is unacceptable to the user if he loses all the work he's done just because he had to stop for a while.

Mark: As much as I hate to admit it, I have to agree with Eric.

Eric: Thanks, Mark.

Linda: The problem is the whole process takes the user from an empty page to a full page step by step. You can quit at any time, but what you end up with is whatever you put in. The program can't add content on its own.

Mark: Can the program check to see if it terminated properly? Is there some sort of flag you could set?

Linda: Yes, sure, but then what would you do?

Mark: Well, if the program discovers an abnormal termination, then we can send a message to the user . . .

Jeanette: I think I see where Mark is heading with this. We can tell the user that he can edit the page or start over again.

Mark: Right. In fact, if he wanted to start over, couldn't you automatically delete the page he didn't finish?

Linda: Yes, but it would be more work . . .

Eric: Do it, but get me an estimate on the work first, so I can let marketing know if there's going to be a slip in the schedule.

And so it went. We solved one issue after another through discourse at our weekly meetings and through e-mail streams during the rest of the week. Not all the meetings were as crucial as the one presented here, but most of them were useful to me. Sometimes, I contributed; at other times, I just took notes.

The other interesting development during the project was the working relationship that Jeanette and I managed to establish. While she concerned herself with the flow of the program, the arrangement of screen elements, and the instructional text that appeared, she recognized the value of the Help file. She listened to my concerns and offered advice when asked. She even taught me about good user interface design. We became a dynamic duo, a one-two punch during the meetings. We fought for the user. We fought for simplicity. We ganged up on Eric or Linda whenever we thought we could win a battle for the user.

The Usability Test

Eventually, I had a complete, working Help file. I had listened carefully to the concerns of the team members, so the Help file reflected the priorities I gleaned from them. The initial help topic from each of the program's screens was introductory, but it contained warnings, notes, and important background information—information the team decided the users had to know. At the bottom of this short introductory topic were all the links to the other relevant information: the procedure, glossary, options, and other explanations.

By this time, the HomePage 2.0 product was ready for beta testing and another round of usability testing. Since this product was for novice computer users, usability testing was vital. How else would the development team learn if it was on target? Jeanette had conducted some usability testing very early on in the project, using mock-ups of each screen, but now it was time to put the actual program to the test, in front of real users.

For this round of testing, the company shipped the product—along with the online help I had created—to another department in the company that specialized in usability testing. Not every company has the luxury of a usability department, but we are one of the few lucky ones. As it turned out, this would be the only round of official usability testing the product would receive.

While the product was away being tested, we all continued to refine our work. Still, emotions were mixed. Since we had met our first deadline, there was at least a brief flash of optimism and even relief. But while everyone stayed busy, the ongoing off-site usability tests did nothing to ease the tension. Jeanette had accompanied the product and was helping to conduct the tests, but reports were few and far between. Linda tried to break up the tension by organizing a voluntary team lunch. Everyone went, but the lunch did nothing to allay our fears. All we talked about was the project.

When Jeanette and the usability test results finally returned, we realized we had been worrying for nothing. Well, almost nothing. Jeanette gathered the project team in the meeting room to review the results and address the issues.

"The usability tests were hugely successful," she began. "Users were able to build their webpages with minimal instruction. They even enjoyed the experience." She paused then, looking around the room. "There are still some problems, though. Users didn't understand how to use some of the features, like building lists and creating links. Since these are vital to most webpages, we have to do a better job of presenting the task, of making it intuitive. We've come up with some ideas . . . " She handed around copies of the usability summary report.

She continued to emphasize the new ideas to solve problems, rather than the problems themselves. There was no blame, only possibilities. Everything she said sounded positive, as if we could make the product even better now that we knew how users react to the product and what they want it to do.

The entire development team got some valuable comments from the testing. I know I got my share. The most valid complaint was that users had to wade through the introductory material every time they opened the Help file. The information they needed was never on the first help topic, even if links to it were. Often, if the users did not see the information they wanted, they just closed the help, frustrated.

The next day, after everyone had time to digest the usability testing results, the development team met again to plan the next step. The mood was upbeat and confident. The team had made its first deadline, delivering a product that exceeded expectations, even if it needed a little tweaking here or there. Linda addressed the team.

"Congratulations, everyone. Now, let's get back to work."

Everyone chuckled.

"The user interface has to change, and Jeanette is working on some of the new ideas. Chuck, work with her to redesign the front end. I specifically want you to keep me updated on the changes to the Building a List screen. Okay?"

"You got it."

"Most of the functionality of the application won't change, and some screens won't even be affected. Others, though, are going to need a lot of work. Keep the usability test report on your desk, and refer to it often. We need to make things easier to use. Questions?"

There were a few; then the meeting ended. All the changes to the user interface directly caused changes to the online help. Now I had my work cut out for me. I had to improve the design of the online help as well as keep up with the changes to the product. I got to work.

The First (and Second) Edit

Based on the usability test results, I determined that the most sought-after information for each of the screens was the procedure. How do users create the desired element, using only their wits and the product's user interface? What questions would arise during such a procedure? This was the information that had to appear on the very first help topic from each screen.

I began editing each of the main help topics so that they displayed the procedure, not the introductory material. Given that all the links were already defined, this was not an easy job. A Help file is like a finely crafted puzzle. When it is complete, all the pieces fit together perfectly. The links are all double-checked. The titles are all accurate. When you begin to rearrange the pieces . . . well, you can imagine the headache you could create if you aren't careful.

During the next few weeks, I made progress. I edited the topics to match the programming and user interface changes. I also made the structural design changes I needed to make. Things were going well. I had finished editing about half the Help file when suddenly I had to stop.

Did the project end? No. Was I fired? No. I had booked time off to attend the annual conference of the Society for Technical Communication (STC). I'm an active member in my local chapter, and this conference promised to be quite fulfilling. Among the many topics to explore at this year's conference were online help solutions. I wanted to know more about helping users navigate through online help. I wanted to know how other people were guiding users to the answers they sought.

I had promised my boss that I could get these answers at the conference. Part of my proposal to her was that the conference would arm me with information and techniques that could shape my current help project. I had no idea how right my proposal would be.

The conference lasted three days, during which time I attended over twenty seminars, presentations, and workshops. The highlight, however, was attending several in which the presenter discussed different views on how people use documentation and online help. These seminars opened my eyes to new possibilities, and I emerged from the conference armed with more insights, new ideas, and renewed energy.

When I returned to work, I immediately set up a meeting with my technical writing peers to validate my ideas on the whiteboard. After some minor modifications—details and exceptions are the last problems to be solved—we agreed the new design would improve the Help file threefold. Excitement raced through my veins as I started anew to edit my Help file. This was why I had become a technical writer!

What was so exciting? What was the change that would make such a difference in the Help file? I'll tell you. I found a simple way to present three topics as one. In other words, when a user opens the Help file, he or she will see one help topic with three parts, but the three parts are three separate topics, all with the same title. Cool, isn't it? Want to know how I did it?

It was actually very simple, although it took a lot of work to implement. I used a hypergraphic, a bitmapped picture that has multiple links. Each picture displayed three tabs, like the tabs of a manila folder. I labeled the tabs Overview, How To, and My Choices. Actually, I used a series of three hypergraphics, one for each of the three topics. Each picture showed a different highlighted tab. The two unhighlighted tabs linked to the other two topics, so users could access any of the three topics with a click of the mouse button. The hypergraphic, then, linked three separate topics so closely that they appeared to be three parts of the same whole.

Help navigation issues have been around for years, but online help is a learned application. The more you use it, the more you understand how to find the information you need. Today, if the target audience is technically savvy, they already know how to use Help files. Novice users, on the other hand, present a special problem. When novice users open an online Help file, they are not familiar with its functional intricacies. They do not know what to expect or what each button might do. They can easily, and quickly, become frustrated.

It is the technical writer's responsibility to solve navigation issues and make the Help file easy to use, especially for novices. The trick I employed put all the pertinent information at the users' fingertips. I still put the procedural information on the first topic displayed (as the How To tab), but now introductory information and navigational clues were also available.

The whole process took some work, since I had to create the hypergraphics and then link them correctly for each three-topic combination. If I thought editing the Help file for the second design was difficult, this work was three times as trying. I had to start from the beginning, editing the topics that I had already fixed. Then I progressed to those topics that still reflected the original faulty design.

The work went along smoothly. The design was the hard part; the implementation was relatively straightforward. Besides, I was driven to finish by the knowledge that I was creating something special, something that would truly meet the needs of its audience. I was creating a Help file that I believed would make the product better, and usability evidence supported this belief. I was creating a Help file that users might really use.

From the company's perspective, the Help file should reduce customer support calls, thus saving untold thousands (or even millions) of dollars. At least that's what I told myself.

The Project Shake-Up

Just as I was finishing this round of edits to the online help, the whole focus of the project changed. Originally, the project was a simple upgrade, improving the functionality and ease of use of the more limited HomePage 1.5 product. Somewhere in the corporate power structure, however, an idea had hatched to make HomePage 2.0 something quite different.

There was also talk that the company was going to cancel the project in midstream. Although it seemed unlikely, especially when you consider how much time and money the company had already invested in the project, anything could happen. We've seen worse. The skeptics believed that we would start from scratch to develop the next phase, whatever it might be. When this type of news hits the floors, it travels faster than the speed of light. Soon, rumors of layoffs and cutbacks reared their ugly heads. Eric called everyone into the meeting room to quell the rampant rumors and explain what would happen to the project. "Good morning, everyone. I have some news . . . "

The room became very quiet.

"I have good news and bad news. The good news is that there are no layoffs, and the project is not being canceled. Let me emphasize how important this project is to the company. They are not going to pull the plug on it just like that." He paused, to gauge the reaction around the room. A few people were visibly relieved; others were waiting for the rest of the news. "The bad news is that the project will be changing direction slightly . . . "

"Slightly?" asked Linda.

"Yes. The company wants to widen its goals to take advantage of growing Internet popularity. Since we have a presence on the web, we want to take advantage of that by hosting our own online homepage community. To achieve this goal, the company needs a method of registering its users' homepages into the community, so we've decided to use HomePage 2.0 as the entry point. It's perfectly suited for this. Our customers already use the product to create their webpages. After that, they will need somewhere to put the new pages, right? It's the perfect tie-in. In fact, HomePage could *encourage* our users to register their homepages with us. It's a brainstorm!" He paused, then added nonchalantly, "We just need to add a few things to the process—particularly at the end."

Eric's announcement was still echoing off the conference room walls when it dawned on me how much work the new plan would involve. I sat there for a moment, silently letting the news sink in. Then I glanced around the room, and I could tell from the faces that everyone was thinking the same thing.

No one spoke. Eric shifted uneasily in his chair and then asked, "Any questions?"

After a moment or two, Linda asked, "So how much time do we have to make all these changes?"

Eric turned the question around. "How much time do you need?"

"First we'll need some requirements details so we can determine the scope of the additional work. If we need to add screens and functionality to the end of the HomePage process, that will take some time. From what you've said, it sounds as if we'll also have to establish and maintain a connection to the company's online community. You have to give us time to work up some estimates. I'll swing by your office with the figures no later than the end of the week. Will that be sufficient?"

"Yes, but don't wait longer than that. The sooner we get started, the better."

The meeting ended and everyone dispersed. Linda grabbed the lead programmers to look over the work.

I wandered back to my cube, deep in thought. I had my own work to worry about. While most of my Help file could remain intact, I would have to add more topics and edit some of the existing ones. I needed to stay on top of the changes, which meant more meetings with QA and development. The whole team now had to put more time and effort into the project, just when everyone was winding down and gearing up for the next project.

Morale of the team took a severe hit. Fortunately, Linda and Eric agreed to give everyone enough time to do the job right. As Jeanette mentioned later, "If we have to put more effort into this project, at least we can do it right." I knew what she meant; no one wants to feel as if she is wasting time. Despite all the problems, everyone on the project still wanted to see it succeed.

At the next status meeting, Linda floated the prospect of a launch party as an added incentive. A launch party can take many forms, but usually it is an off-site lunch when the product finally goes out the door. Sometimes the company also offers T-shirts emblazoned with the product's name. Both forms of bribery can be effective incentives to an overworked team.

Back to Work

The decision to marry HomePage 2.0 with the online community hampered my efforts to deliver my online Help file. How? For starters, I would have to document the new screens and the new functionality. That was a given. But there was more.

The decision changed my target audience. In the original project plan, 99 percent of all HomePage users would be computer novices. Now, computer experts would have to use the program to register their pages in the online community. Suddenly, a whole new set of users might access my Help file to discover how to do an unfamiliar task.

I needed to review my Help file structure and content to determine how best to serve all possible audiences. The file should not get in the way as users search for the information they need, no matter what their level of expertise.

I began again to edit my Help file from the start. This time, I looked for phrases or definitions that expert users might construe as condescending. I reworded these phrases without damaging the friendly tone of the Help file. Next, I took all the common definitions out of the text and replaced them with a link to a pop-up glossary definition. This step even made the Help file more effective. Procedures and explanatory text were shorter.

By the time I got to the last topic, the new screens were ready for me to preview, test, and document. This part took quite a bit of effort, since I had to interview Linda's team of programmers. I contacted Chuck (the UI programmer) about the new screens, and he told me to come by his cube whenever I had a question.

I decided to write down the questions I had before my visit. I didn't want to keep interrupting him because he had other work to do. With questions written down, I could zip through them quickly. If another topic arose during our conversation, I could pursue it without worrying about forgetting my original set of questions.

I knocked on Chuck's cubicle wall. "Do you have a few minutes to answer my questions?"

Chuck turned and said, "Hello, Mark. Hold on; I was just finishing up a bug fix."

I cleaned off a chair and sat, organizing my notes while Chuck typed away, his attention on his monitor. After a minute, he saved his work and swung his chair around to face me. "What's up?" he asked.

"I've got some questions about how the new features work. For example, what happens if the user declines to register?"

"That's easy. When the user finishes building a page, the program asks him if he wants to register his page in the online community. If he says yes, then he's connected to the host where he can select the community he wants to be part of . . . "

"Like what?"

"Well, if his page is about sports, there's a sports community. If the page is about baby pictures, there's a family community. It's pretty straightforward. The host UI guides him through the process."

"Okay. What happens if the user doesn't want to register?"

"The program then switches to the old Congratulations screen. From there, all he can do is save the page to his hard drive."

"I see. Thanks. What happens if the user gets timed out during the registration process?"

"That's a problem we've struggled with, since the host connection is out of our control. We finally decided that if the user is disconnected, he has to start over. He is not registered until he completes the process."

"What error messages are likely to appear?"

"Oh, you know, the normal sort of connection errors, generated by the communications software."

"Really? I've seen some of those cryptic messages. Can we trap the messages and provide our own, more understandable text?"

"Well, yes, I suppose we could . . . "

"Great, thank you. I can write clear error messages if you provide the old messages and what they mean."

"Sure. We can do that."

I thanked Chuck again and returned to my cubicle to continue my editing work. He helped me understand the process, and he allowed me to make the project better. Clear error messages might be a small step for the project, but it represents a huge leap for technical communicators. The program is better with me on the team than without me. I make a difference.

A Gold Master!

Eventually, the QA department passed the program and my revised Help file. The company declared the product a "gold master." In corporate slang, we said the product has "GM'ed." This term is not to be confused with a product being "GE'ed," which means it was "good enough."

Everyone connected with the project attended the launch party, a lunch at a swank seafood restaurant on a beautiful late summer day. After we congratulated each other, we ate and drank and discussed anything other than the project we had just completed. During dessert, the project manager gave a short speech and presented little gifts to almost everybody. Then we all raised our glasses and cheered.

Such is life in my company. Not all corporations work this way, and not all projects develop in this manner, but this was a true experience. I learned quite a bit during the project, and as far as I can tell, the product has been a success.

If you're wondering what happened to my Help file, it's out there with the program, being used by (or at least accessible to) millions of people. How the users have received it, I cannot say. I get virtually no response from the users in the field. However, this I do know: the program is being marketed worldwide, so my Help file is being translated into other languages. So far—and this is the truth—the Help file has been "translated" into Australian English, Canadian English, and UK English. I know because I've recompiled the Help file with the translated source files.

Technical writing is not always this dramatic, but since I love doing what I do as a technical communicator, it's always this much fun. Of course, being part of a good company and part of a superb team are always contributing factors to the level of fun you have at any job. I wouldn't trade places with anyone.

Tim Casady

Tech Writing and the Art of Laziness

A graduate of the English department at Briar Cliff College, Tim Casady currently lives in Chicagoland. As a writer who has consulted for many different companies, Tim knows the importance of maintaining a sense of humor.

"Here's the user documentation for the previous version of the software," says my new client manager, pointing to a shelf. Seven thick binders weigh down the shelf. "Look them over," my manager says, as he retreats toward his office. "They can be used as a starting point for the new documentation."

In two trips, I lug the manuals back to my new cubicle and throw them down onto my desk. Then I go to the break room and come back with a full cup of hot coffee. I don't normally drink coffee, but I'll need it today. I plop down into my chair and stare at the manuals. I turn to my computer monitor and look online at the application that I'm supposed to document. With a click of the mouse, I bring up the account screen.

I try to pretend that I am an end user. If I were just looking for information on how to use the account screen, where would I find it? The menu option for online help is disabled. The seven paper manuals all have fairly nondescript titles, so it's not obvious which manual I should start with. I pull out a manual called *Operations* and flip through the table of contents. Then I flip through the index. I can't find what I'm looking for. I begin the process with the second manual. Then the third. Then the fourth. I do this to all seven manuals, and I still can't find what I'm looking for. I start flipping through the pages of the manuals at random. Finally, I find a section on using the account screen. Relieved, I decide to take a sip of coffee. The coffee is now cold. So much for user friendliness.

> *Knowing what to include in the documentation is important. But knowing what not to include is equally important. This discernment is essential for a tech writer. However, this task is not easy because you must constantly rely on people who have different opinions on what's important.*

The account screen is one of the most common screens that the users access, but it's also one of the most difficult screens. If I, as the technical writer, have difficulty finding such basic information in the manuals, what hope has the user in the field? Surely, the users would like to get their job done with the least amount of hassle possible. They would like time to drink their coffee and not spend all afternoon flipping through the manuals. I understand the laziness of the user in the field, and I want to develop my documentation around that laziness.

The next afternoon, I'm sitting at my computer screen. I'm developing an outline for a quick reference guide. This manual would help users complete their most common tasks with the least amount of page flipping. Such manuals have been successful on other projects, and it would probably be successful on this project as well. I save the outline and then create a new file. I would like to outline what information would be best to include in an online help system. Why make the user search around for a missing manual to find a field description that they could get with a click of the mouse?

Someone knocks on the wall of my cubicle. I look up. A developer stands there, with a doughnut in one hand and a two-inch-thick binder in the other hand. He didn't come here to offer me the doughnut.

"I heard you were revising the user documentation," he says, handing me the binder. "Here's some material that will be useful to you." The binder is so full that some pages are tearing out.

"Thank you," I say. "I'll look it over."

"It's all important," he says, as he retreats to the safety of his own cubicle.

I open the binder to the first page. On it is a cramped diagram. It could be a flow chart. It could be an org chart. But it is such a badly designed diagram with acronyms in such tiny fonts that I can't determine what it is. I'll have to get clarification from the developer about it. I flip through the next few pages. There are more diagrams, obviously designed by people who don't know how to design diagrams. Then I get into the main body of the document. I see information on internal variables settings and function call information. Why did the developer give this to me? This document is page after page of pseudo-code, describing the programming logic that goes on behind the scenes of the application.

I estimate that 98.267 percent of this material will be useless to the user in the field. But I decide to look it over more carefully. I do. I was wrong. Actually, 99.267 percent of this material will be useless to the user.

Knowing what to include in the documentation is important. But knowing what not to include is equally important. This discernment is essential for a tech writer. However, this task is not easy because you must constantly rely on people who have different opinions on what's important.

My manager stops by my cube to see how things are going.

"Hey, who's responsible for the Config subsystem?" I ask. "The user doesn't have to set up this subsystem at all. There's a lot in the old user documentation about this subsystem that probably doesn't need to be there."

"Carol used to be responsible for that, but she left five months ago," my manager says. "Just go ahead and leave that information in the documentation."

I have to deal with that attitude a lot. "But users can really mess up their application if they use this subsystem; it's all correctly configured automatically at install, and users could just create some headaches for tech support." I then give him some specific examples of past tech support problems created by users modifying the Config subsystem.

He shakes his head. "Go ahead and do what you think best."

As my manager heads off to his meeting, I know what I think is best—to remove what the user doesn't need, to remove what would get in the way of what the user needs, and to remove whatever makes the user feel that using the manual is futile.

Although I am adding a quick reference guide, I plan to actually reduce the number of manuals by more than half and reduce the overall page count. Computer manuals shouldn't be judged by the same standard as gold. The more your gold weighs, the more it's worth. But just because one documentation set weighs more than another documentation set doesn't mean that it's worth more. In fact, the more your documentation set weighs, the more likely that it's worthless. The more manuals the user has, the more manuals there are to get lost. It's easier to keep track of one or two manuals than it is to keep track of seven.

At the end of the week, I give a copy of the documentation set outline to my manager and other team leaders. This outline includes not only the proposed quick reference guide and the online help but also my ideas for revamping the existing manuals. However, I keep this outline under ten pages. This is because I understand the principle of laziness in my managers and coworkers. If I go to someone's desk and leave five hundred pages for the person to review, I'm likely to receive no useful review comments whatsoever. This thick manual will be buried somewhere on the desk and forgotten. However, if I hand out a document of ten pages or less, I'm much more likely to get useful review comments.

In school, students hated outlines, viewing them as extra work. But in the tech writing world, outlines save you a lot of extra work. Outlines allow your manager to see what you're thinking of doing before you actually do it. If your manager hates the outline for a manual, you haven't wasted that much time. But if you actually write the entire manual, and then find out that your manager hates it, you've wasted a lot of time. Therefore, outlines can save you literally weeks or months of unnecessary work.

Three months later, my documentation set is complete and sent off with the release. Since I was brought in for only this one project, I will probably be finishing with this client shortly and sent off to a different project somewhere else. But my manager calls me into his office. "I'd like to use you again on a future project if possible. You've done a lot of hard work on this project."

"Thank you," I say, knowing that I performed a lot of hard work on the project. But I also brought a lot to the project by understanding laziness.

Julie S. Hile

I've Been Working on the Railroad

Re-Vision at BNP Railway

Julie S. Hile, principal of The Hile Group, assists public and private-sector organizations with performance improvement interventions, especially in continuous learning, workplace communications that maximize associate involvement, and safety and health. Recent customers include American Commercial Lines LLC, Amtrak (National Railroad Passenger Corporation), Burlington Northern Santa Fe Railroad, CSX Transportation, Crowley Marine Services, SeaRiver Maritime, and Kansas City Southern Railroad.

Julie is currently thinking and writing about extrinsic motivators and about organizational documents as performance interventions. She and her associates are researching employee turnover in the marine industry. Finally, she is perpetually refining her work with the transfer of learning.

Julie has been a member of American Society for Training and Development (ASTD) since she entered the field in 1989. She has given many ASTD presentations over the years. In September 1999, she collaborated with Dr. Mary Broad to present "A Transfer of Learning: Ensuring Organizational Payoff Back on the Job" for CIC-ASTD. Hile has served in various capacities on ASTD boards for six of the past nine years. Julie's other professional affiliations include the American Society of Safety Engineers, Awareness to Action, and Businesses for Social Responsibility.

Julie and her partner, Bob Broad, enjoy the company of their son, Dylan, eight, and their daughter, Rachel, five. When not working for pay, she loves to cook, run with the family pooch, sing and play the piano, garden, read, and work for social justice.

My relationship with workplace documentation was coming unglued. What innumerable other texts had I blindly and ineffectually penned, not realizing that the words inside needed to be the words of the people whom they would guide? What processes of negotiation and clarification had I short-circuited by coming on the scene as the "writing specialist," the "language woman"?

As the cab rounded the bend in the beltway, the company name—in immense fluorescent blue—loomed into view. It spanned the three top stories of the tallest building in the city center: BNP RAILWAY. Whooa, I thought to myself. Julie, how on earth do you get yourself into these situations? I recollected memories of the buildup to this opportunity: the paperless payroll documentation and implementation design. Intense conversations with Belle Washburn, the administrative assistant on the payroll project, about her friend, Toby, who headed up the safety department. "Boy, could he use some of what you're selling," she'd said, shaking her head. The safety committee member communications program and train-the-trainer. And then Toby, on the conference call two months before, saying, "How about you come down here, Julie, and help us with a full staff meeting? I think we need you to help us get this rule book on the road."

"BNP is pretty big stuff around here, huh?" I ventured at the back of the cabby's head.

"Yes, ma'am. One of the biggest games in town," he nodded, turning onto Water Street, drifting into the left-hand lane. "Great company to work for. You here with BNP?"

"Uhh, yeah. I'm a contractor. A sort of teacher–writer."

"Oh. You an independent?"

"That's right. I'm helping the railroad rewrite its safety rule book." End of conversation. The driver was pulling Billie Holiday from the radio dial, eyes still on his driving. I stretched my arms around to rub away the chill, felt the baby stir beneath my ribs, took in the streetlight glancing off of magnolias, breathed deep as we pulled in to the courtesy drive of the Ponce de León Hotel.

The following morning I was up early and in the meeting room, getting tables, chairs, and flip charts set up. The task with which I was charged was to lead the revision of the company's twenty-year-old safety rule book. One might think that a rule book revision process would begin in a small, quiet room with a copy of the old book and an expectant PC screen. But I had come to see, in collaboration with Jim Sotz and Cindy Haver (my two lead contacts on the project), that we were about to not only change the words in the smallish yellow book that all BNP employees were required to carry, but also about to revolutionize the way the railroad's people understood safety. The book—which held 930 rules that had accumulated for more than 150 years, arranged under dozens of convoluted subheads—would, by the end of this week at the languorous, salmon-stuccoed Ponce de León . . . implode.

Cindy's drawl rounded the corner before her. "Well, we've got them all here now. Let go of it, Yoda!" She enveloped me in a hug. "Jim here is worried that we've got overkill, but I say what the hay, we're on our way. Hello, darlin'." She scanned the room. "Whooee, I see you're already hard at it."

The room was looking good: markers on each table, water pitchers and glasses at the ready, eight-top round tables draped in white cloths.

"Hi, Julie. How's that baby?" Jim offered his gentle squeeze. "What is it now, seven months?"

I gave him the pregnancy update: feeling fine, baby moving, not so tired anymore. We settled, the three-plus of us, at a side table and began to review our strategy.

My original recommendation, months back, had been that I would review the rule-laden "Old Yeller" and meet with Jim and Cindy and maybe a couple of guys from the field. I'd draft up a new book. So, for example, if I were revising one of the current safety rules about switch and derail operations, I might write: "When working around a power-operated switch, a moveable point frog, a derail, or its connection, keep all body parts, clothing, and tools clear of all moving parts, unless the open points have been blocked or absolute protection is provided." We'd meet with a larger group to review the draft. I'd revise and polish. Then we would go to print.

Then Jim had started with his questions.

"What kind of a book do we want this to be?"

"What kind of a book do our people want this to be?"

"What is a safety rule, anyway?"

"What are we going to do about the rules in Old Yeller that people can't comply with? the ones that force noncompliance?"

Cindy joined in. "And what about the dadgummed 'gotchas' that stick people like a goat tied to a post? This book is full of rules that are meant to make sure the company has a reason not to do right by a person who has done nothing wrong."

They had a point. They had a lot of points.

"Well," I'd offered, "the questions you're asking aren't questions we can answer. These are philosophical questions. This isn't a conversation about writing a book. What we're having right now is a conversation about the ethics and the workings of authority in your safety program."

Sotz and Haver were listening.

"You're absolutely right to ask, 'What is a safety rule?' " I continued. "I would add, who gets to decide which work procedures are rules and which aren't? What is the difference between a work procedure and a rule, anyway? And a federal or state regulation? And a policy? Where do training and continuous improvement fit in? How much or how little information should be included in something you're calling a safety rule book? And who's to say that printing information in a book you make people carry around with them (mainly, as you have said, so they know which rules they'll have to defend themselves against if they're involved in an incident) is the best way to help them work safely anyway? Maybe that's the worst approach you can take."

"*We* can take." From Cindy.

"What?" I blinked.

Jim and Cindy looked at one another, grins spreading. Jim winked a small wink. "Cindy," he said, "I think she's beginning to get the picture."

So there I was, a technical writer, sitting at the Ponce de León, laying the last of a plan for a day, a week, that would not involve my brandishing a pen and coming up with the language that was going to fill this book. Rather, I was preparing to lay my pen aside in favor of working this purposefully unwieldy group toward dialogue and consensus. We were about to converse. Debate. Convert. Catalyze. I had absolutely no idea what was going to happen.

We expected eighty people in the session: the entire safety staff (including two certified safety professionals and an ergonomist), leaders from four different unions, employees from five railroad crafts. Corporate lawyers. Representatives from the North, South, East, and West. Veteran railroaders and rookies. People from corporate communications. A couple of Toby Helgeson's colleagues from the company's executive leadership team.

Participants began to filter into the room. Jim, Cindy, and I moved to greet people, to make connections. I introduced myself and responded most often, "Yes, this is my first baby. I feel great."

At 7:30 A.M. sharp (a late start for railroaders), Toby kicked the meeting off. On his watch, the company's safety record had seen improvements that were nothing less than miraculous. The railroad's Frequency Index, the number of recordable injuries per 100,000 person hours worked, had plummeted. The number of unsolicited reports of near misses or very close calls had risen sharply, as had the number of employees participating in the studies of those incidents to learn what had almost gone wrong so that the company could broadcast tips for preventing such risks in the future. Housekeeping at train yards and terminals across the property had shifted so dramatically that in some locations, as Toby pointed out to the group, landscaping was actually being done and employee-created murals with spirited safety themes were being painted on the outbuilding walls. Generally, employees practically everywhere had caught the safety religion and were living it like some sort of gospel. Individuals and teams from BNP were receiving national freight rail safety awards by the score. Toby applauded his safety rule book team as he reviewed these many successes. He acknowledged their remarkable work both of late and throughout their careers. He mused, "Here at BNP, where safety is concerned, we've reached the place where every one of us knows in our hearts what is the right thing to do."

He pushed hard at us, challenging us to bring the new, employee-driven thinking into the safety rule book.

"I'd like to see a rule book that is precise, unambiguous, and consistent with the cultural change that has taken place at BNP: respecting employees and holding the quality and sanctity of their lives above all else."

He went on to outline the plan for the week and its projected outcomes.

"By the time we all leave the Ponce de León," he said, "we will have decided just what we want in this new rule book. We will know what the book will look like, which of the existing rules will stay, and which will go. And we will have drafted a new Safety Statement."

Toby turned to me with a meaningful look, that it's-your-turn-Jules look that had become so familiar in the months we'd worked together. My eyebrows lifted involuntarily, as if to hoist me out of my seat and into the fray. Okay, I thought, we go.

A quick introduction and safety briefing, and then I hit the group with a question: "How would you like BNP's next safety rule book to look? How would it be similar to and different from today's book?"

Participants talked for a short time and shared their responses with the large group. I noted their comments on flip charts located variously around the room: pocket-sized, craft-specific, lots fewer rules, more rules, fewer rules, a compendium of anything a person would ever want to know about safety, a job tool, exactly the same book we have right now, period. A pause.

The list got longer. And broader.

I began to see people's passions around the current book. Lawyers who had won settlements on the discrete strengths of a particular rule. New employees who were overwhelmed by the wordiness, the obtuseness, the stark combativeness of Old Yeller. People who had helped author rules after seeing close friends horribly disfigured, and worse, on the job, who felt that removing those rules was an act of irresponsibility and calamity.

More and more, the cover and pages of the little volume came to life for me. My relationship with workplace documentation was coming unglued. What innumerable other texts had I blindly and ineffectually penned, not realizing that the words inside needed to be the words of the people whom they would guide? What processes of negotiation and clarification had I short-circuited by coming on the scene as the "writing specialist," the "language woman"?

I glanced over at Jim and Cindy, searching for some sign of approval, disapproval, glee, terror. Their faces were poker-blank. They knew that we had reached a critical juncture, that their positions had to be perceived by key resisters as neutral, in process. No help for me there.

The notes I was making on the flip charts increasingly moved to capture the gaps, the critical questions, the character of the discord in the group—my way into the complexity of the thinking that was going on. The list drew others in the group toward common ground, too, if not to consensus. This was hard work.

At noontime I sat down to rest. Flip chart pages covered the walls. People milled around, profoundly agitated, concerned. Some angry. I'd not written a single rule. What was I doing anyway?

I stood, breathed deep. Walked away from the group. Away from the room. Away from the Ponce. Thinking. Thinking. We needed a template. The idea began to dawn. What was really going on here was a struggle to sort a tremendous amount of valuable, credible information about railroad safety into various genres. And the genres were defined not so much by form, though that was relevant, too, as by relative mandatoriness. That is, was the information absolutely critical to a person's safety? Did we want to require compliance to a given rule or only strongly recommend it? Was the information really about training, about learning? Was the information, in fact, about safety at all? When I looked up from my reverie, I was greeted by Spanish moss and the *creee* of tree frogs. I'd traversed the golf course adjacent to the Ponce and would have to hustle to get back to my room and my laptop in time to write. For, yes, now I was ready to write.

I returned to the meeting room fifteen minutes later with a decision-making flow chart that would organize our thinking for the rest of the day—indeed, for the rest of the project. The chart asked, simply, Is the information you're sorting safety-related? If yes, is it critical to safe job performance? If yes, is it applicable to all crafts? If yes, can all affected employees comply with it? If yes, can it in all cases be enforced? A yes answer classified that information as a core safety rule, one of those rarities that would establish the basic safety principles for BNP. No answers led a piece of information to alternative destinies: craft-specific safety rules, craft-specific recommended work practices, training materials, regulatory materials, and policy or audit materials.

I used the switch and derail rule I'd at first planned to rewrite unilaterally as a test run for the flow chart. It came through a large group discussion something like this:

> *Is the information you're sorting safety-related? Yes, keeping body parts, clothing, and tools clear of all moving parts near a power-operated switch, a frog, a derail, or its connection is safety-related. If yes, is it critical to safe job performance? Yes, keeping clear is critical to safety.*

But how effective is a rule that tells people not to stick things into parts of machines that could pull bodies in or chew bodies up or turn tools into projectiles or break if so fed? Do we really want to tell people not to do so basic a thing? If we mandate so puny an action, then how microscopically do we need to dictate job procedures from every corner of our operations? And how well-advised is the blocking of open points? Is that a practice we want to effectively legislate out in the field?

We paused altogether; the enormousness of the task we were defining shone brighter by the moment. I squinted my eyes, shook my head, felt the baby roll. My hand straightened from pen to karate-chop and pressed in slightly above my waist to prevent Baby from connecting a sharp-heeled wallop to my ribs. I grinned to feel the expected jab to my knuckles. Ha. Gotcha, I thought. Cindy winked at me from a few feet away and mouthed, "We're in ever-loving labor! BNP Railroad, I mean."

On we went. Will writing such a rule into a book cause people to keep clear? Or is this rule really a "gotcha"? In other words, does the language function in the main to litigate against people after injury, rather than promote a critically engaged, effectively trained and supported, hazard-zapping safety culture? Is this information better treated as training or safety briefing material—with perhaps something about keeping clear in general in the core rules?

If yes, is this rule applicable to all crafts? No. Switch, frog, and derail operations are not applicable to all crafts. Only transportation and mechanical employees handle those operations. So at least part of this information can only be presented in the craft-specific rules and recommended procedures. But yes, keeping appendages, bodily or otherwise, clear of the moving parts of any working equipment is indeed a cross-craft concern. Again, might we have core rule material here?

If yes, can all employees affected comply with it? If yes, can it in all cases be enforced? Does it offer "absolute protection"—as defined by whom? Isn't protection only absolute if it prevents an incident? What does *absolute protection* mean? And if we can't define the term clearly for application in all circumstances, then how can we know (after the fact) whether a person has complied with the rule or not? More important, how does this information (before the fact) help a person know what is the right thing to do?

The flip charts were accumulating as we struggled for clarity. John Maser, a gracious, graying, gently pink-cheeked railroad lifer who had written dozens of highly prescriptive rules to help BNP learn from its missteps, took the floor. "I understand what you people are doing here. I see that your definition of *rule* bars many of the current rules from legitimacy. We can't enforce them, you say. We can't comply with them. Well, what I say is that if we choose this path, I'll sell every share of my BNP stock because they won't be worth the paper they're written on."

From across the room, quietly, came, "I'll buy every one of them. Happy to." Marv Shea, an old friend and colleague of Maser's, stood there in open challenge. "And when I die, I'll be a far richer man than I am today, whether I'm counting the dollars in my bank account or the points laid by for me in Heaven because I'm doing the right thing here today."

Turning point, I thought to myself. Turning point. Is it going to turn? Maser excused himself from the meeting. I asked whether the group needed a break. "No need," was the communal response, "Let's move it on."

We crafted a core rule: "You must be qualified to operate tools and equipment." And another, "Inspect all tools and equipment and related safety devices for unsafe conditions before use, removing them from service if defective." A third read, "Use the proper tools for the purpose designed. Unauthorized modifications are prohibited." We talked about the value of teaching people to see the hazards in any work situation, to assess the options for safely accomplishing the job, and to ask for help from their colleagues when they harbored even the slightest doubt about how to safely proceed.

My markers were low on ink by the time we'd moved on through a trim section of craft-specific rules and recommended practices. Meeting participants here and there had risen to the flip charts we'd placed near them and begun drafting phrases. Jim and Cindy noted this development with knowing flashes of eye contact: we were all getting the hang of it. We were learning for the first time—despite the group's 2,400 or more years of collective experience in writing rules on the railroad—what a safety rule could be, should be, must be in order for people to consistently go home to their families healthy and whole.

We agreed to divide the group into satellites in order to tackle the earnest drafting of rules. Each group included representatives of all crafts and levels of authority, with legal, communications, and corporate participants checking in from group to group. My job was indeed floating among groups, not writing myself, but teaching writing. This was exhilarating work! I saw this BNP team learning the safety rules from the inside out through a process that embodied all that Toby and Jim and Cindy had been working for during the previous eighteen months. The sorting and writing became a mechanism through which people could decide what their colleagues needed to be required to do, what it was impossible to require them to do, what it was insulting and disrespectful to require them to do. Even as the satellite group recorders committed these decisions to the page, I could see a growing confidence in the experience, the competence, the wisdom, and the commitment of BNP employees systemwide.

We worked at fever pitch through lunch and on into the late afternoon. To close the session, Jim, Cindy, and I configured one last set of writing groups, this one to draft an overarching safety statement for the effort. We planned to define cross-functional groups so that all voices would be equally represented and, once the groups had drafted their paragraphs, to pull the best from all into one excellent statement.

Our participants mutinied. No one was willing to budge from the groups they'd been writing in all day. I gulped. Jim got quiet, formulating his response. Cindy paused, swept a glance across the folded arms and stiffening backs before her.

"Well, let me see, I guess this would be one of those deals where the job procedure as we have defined it is not critical to safety. Looks like you all think there is a better, safer way to write a safety statement."

Silence.

"What the hay. Go to it then. Let's see what you can do."

One group in particular had me concerned. Most of the craft-specific guys had clustered together, and they headed into the Spanish Room. I hoofed it after them, arriving just in time for them to turn, shake a finger at me, say "Out!" with a grin, and firmly close the heavy door behind them.

Nuts. What in the world would these yahoos come up with? Their writing had been solid enough through the day, but I worried that they'd throw in some sort of prounion language or that they wouldn't be able to articulate an issue of such import as the safety statement. I didn't want them to look less capable than the others. Fingers drumming on my cheek, I turned to find Jim standing right behind me. Without a word, we shrugged and went to check in with the other groups.

At fifteen-minute intervals, I returned to the Spanish Room, each time to be sternly denied admittance. I was genuinely concerned. After all, I was the tech writer here. I was the person who was supposed to be driving this project. What was Toby going to say? Was I doing my job?

Four o'clock. All doors opened to smoke and loo breaks. The craft group's Jack Carpenter was very, very excited about the draft they had written. When we reconvened, I asked whether he would like to present to the large group first. Might as well get this over with, find out what we'd gotten ourselves into now.

"Happy to." He grinned at Marv and the rest of us. And he read the following:

> *BNP is committed to being the safety railroad in the country. By empowering everyone with the right, the responsibility, and the resources to make safe decisions, we will accomplish our goals. Our ultimate goal is to prevent all personal injuries.*
>
> *This book contains specific, general, and departmental rules governing our work activities. Compliance with these rules will prevent personal injuries.*
>
> *Rules cannot be written to cover everything we do on the job. Therefore, we are empowered to make decisions and take action necessary to prevent personal injuries.*
>
> *Where no specific rule applies, we must rely on good judgment, following the safest course available. We may have to contact a coworker, supervisor, or manual for guidance. No action should be taken until we are fully aware of the hazards involved and have a plan to avoid injury.*
>
> *Remember: No job is so important, no service so urgent that we cannot take time to perform all work safely.*

A long silence followed Jack's last words—so long that he dropped his eyes to the rose-and-turquoise carpet, face flushing.

"Well, I'll be damned," Toby muttered then, under his breath, rising. "I'll be damned. That is one fine piece of work, boys," he offered. "Puts my thinking to shame."

"No need for us to read our draft" came other voices from the floor. Unanimously.

"I guess you showed us what you can do," said Cindy.

"I guess so," said Jim. Hands on hips, looking long at his friends and colleagues, Jim blinked, swallowed hard, made a quick nod. There was not a lot more for him to say. This was an emotional moment.

I looked over at Jack's group. Mel, Cal, Paul, Rory, David, Jigger, and Steve knew that they had accomplished the extraordinary. Paul, who had keyed the text into his laptop computer as his colleagues had crafted it, snapped the computer shut and slapped Jigger on the back. Laughter swelled across the room and a general celebration ensued. What a day.

I packed up my markers and rolled up page after flip chart page for their transport home to Illinois, where I would clean up the writing and begin formatting a draft for further review by the large group.

I thought—Jab! Darn, Baby got me!—that my work as a technical writer for this project left me both more and less important than I'd been before in my capacity as resident writer. More important in that I had grown to be responsible for monitoring and making global sense of the collaborative process as it evolved organically in the uncanny grassroots manner that we'd experienced in the meeting. Less important in that I was no longer actually doing very much of the writing at all. In fact, I was left helping the group sweat through details like tone, consistency, verb tense, and formatting rather than sticking my pen in there where BNP-ers were very competently writing away.

A quick hug from Toby. A "you take care now, Girl" and a peck on the cheek from Cindy. A gentle squeeze on the arms from Jim. And I was out of the Ponce, back in a cab, breathing deeply to memorize that remarkable magnolia scent. Going home.

It's worth noting, I think, that the safety statement Jack and his BNP craft-colleagues wrote in a single draft there in the Spanish Room at the Ponce de León has migrated across the entire U.S. and Canadian freight rail industry. If you pick up any safety rule book today, large or small, micro-managing or empowering, you will see large chunks of their actual language, and you will feel in varying degrees the collaborative spirit of the BNP effort. Furthermore, while use of the participative process we created at BNP has ebbed and flowed within the transportation industry at large, the Federal Railway Administration is currently researching the model's impact on performance with the prospect of endorsing it as an industry standard.

Shawn S. Staley

Daze at the Round Table

Shawn S. Staley grew up in a small Midwestern farming community. Her education credits include a B.A. in English and M.S. in professional writing. Since college, Staley has worked as the sole technical writer and resident jack-of-all-trades for multiple start-up software companies in the southern United States. She has worked in both permanent and contract technical writing environments. The software industries she has documented to date include sales force automation, restaurant table management, emergency room patient management, and commercial real estate. When away from her cubicle, Staley engages in competitive cycling and dog sports.

A head suddenly and swiftly popped into my cube, and my manager asked, "Ready for our 9:00 meeting?" I turned to face him and concealed a smirk. This was not the first time I had no advance notice or background information about a meeting. In fact, had I ever been notified ahead of time for one of these things? I sighed and made the only appropriate reply: "Yes." I turned to my computer, clicked the Save icon on my Microsoft Word document, grabbed a notebook and pen, and hurried to the conference room. When I arrived, the room was dark. I noticed a figure slumped at the long cherry wood table, sipping pensively from a steaming coffee mug. I knew it was Hugh. He was always on time, and he was a programmer, one of the mole people. I didn't see much of Hugh unless we attended meetings together. The programmers were separated from the rest of the company in their own private, locked suite of the office building. I knew they worked in the dark and liked it that way; they claimed the fluorescent lights caused a computer screen glare that damaged their eyes, so they worked solely by the glow of their computer screens.

None of the programmers understood how the software product worked as an integrated whole. Each programmer worked on specific features of, or enhancements to, the software product and concentrated only on the coding that concerned them. Additionally, they were isolated from the user. They never visited client sites and had never seen how the software was used on a daily basis.

I reached behind the conference room door and groped for the light switch, flipped it on, and selected a seat next to Hugh. We waited in subdued silence as other attendees began to trickle in.

Patty clunked into the room, trying to balance her six-foot frame on high heels. She placed her pocketbook on the table and promptly clunked out of the room for what we all knew would be an indefinite period of time. Marie arrived shortly thereafter with a thirty-two-ounce Dickey's Barbecue cup brimming with water, as well as her standard breakfast of an apple and banana. She was skinny, healthy, and resented. Terry shuffled into the room and popped the aluminum top on his morning can of Coke. His eyes were red, and everyone already knew from the gossipy receptionist that he had been working on a technological problem at the office until 2:00 A.M., a problem he had not yet successfully solved.

I stared absently at my right foot for a while as it brushed against the table leg, then glanced at my watch. I had lost another twenty minutes of work time that I'd never get back, waiting for this unplanned meeting. And where, oh where, was the manager who had popped into my cube to notify me of this meeting, anyway? I tried to subdue my anger, finally reminding myself that I was silly to expect him to be on time and organized.

Finally, my manager appeared with his burgundy Daytimer and laptop computer, which he placed at the front of the room. I noticed that his hair was still wet. After twenty minutes of figuring out how to connect his laptop to the network cable outlet in the wall, the meeting commenced.

"Good morning, everyone," my boss smiled. Only a director of marketing would be cheery and outgoing at 9:24 A.M. "Since we're releasing the product in three weeks, that's fifteen working days, I need to get a quick status report from each of you. We'll go around the room, starting with you, Hugh."

Hugh shifted slightly in his chair. He began mumbling his report, and I couldn't hear him clearly but didn't really care. I was next in line and had to decide what I was going to say and how I was going to say it. Would I be tactful and diplomatic, or honest and scathing today? It would not be pretty, I feared.

When my manager nodded at me, I decided to be honest. Brutally honest. "Quite frankly," I began, "I don't think there's going to be a software release in three weeks . . . at least, not one that includes accurate or complete documentation." I had learned from prior chaotic releases that honesty was the best policy. Otherwise, I'd work sixteen- to eighteen-hour days (including weekends) to try to finish the documentation, only to find that a quality assurance specialist would discover a major bug in the programmers' code that would delay the release for months and require documentation revisions of epic proportions.

I had my manager's undivided attention.

"The programmers promised me a visual freeze of the software one month ago," I continued, "then postponed it by one week, then two." I seethed with frustration but tried to control my shaking voice so I could respectfully defend myself. "Anyway, time's up, and the programmers are still making changes to the user interface. Furthermore, none of them tell me what they're changing or when or why. I just happen to be very familiar with the product, enough to know what I've documented and what I haven't."

My manager's eyebrows formed a V and I knew I was about to be interrup—

"Why," he interrupted, "didn't you tell me about this sooner?"

I shook my head in defiance and gave my usual reply to Mr. Social Butterfly, the manager who claimed to be there when the marketing team needed him but was never truly available for consultation. When he wasn't on the phone, which was rare, his office door was closed . . . sometimes both. "I tried," I insisted, "but you didn't have the time."

"I'm never too busy for my people—you know that," he stated, and there was no mistaking that he was deluded.

"I sent you six e-mails regarding the visual freeze," I said hotly, "and I even called your voice mail three weeks ago and asked you to check your e-mail from me." I felt smug. I knew it, and so did everybody else in the room. I prided myself on doing my job right and sticking up for myself, but right now I just wanted to lack conscientiousness like Patty and ditch this worthless meeting.

My manager nodded sadly and decided it was time to own up to his mistake. He didn't have much choice. He sat down at his laptop and began composing a memo of follow-up information that he would never use. While typing, he began brainstorming aloud. "So we can just have the developers send you a list of every change they've made to the product in the last month."

Hugh sat up straight and began to whine, as most programmers do when the threat of being held accountable for their work is pending. "We don't know what we've changed since a month ago," he protested.

My manager pounced on the opportunity to place blame on the programmers. "Well, Hugh, you might want to start documenting changes, since that is part of your job."

Hugh smiled, which he could do since he was not in the marketing department and didn't have to comply with my manager's demands. "The programmers don't keep up on development documentation. We hate writing."

My manager's eyes narrowed, and he said, "Well, perhaps you'll learn to love it since I'll work with your manager to ensure that you help marketing get the documentation done on time." Hugh folded his arms across his chest and returned to the slumped position in his chair.

My manager turned to me briefly and I nodded my approval, but it was clear to me that I could not contest his decision. I had learned to express myself honestly, but I had also learned that I could only go so far with my boss. I could voice my concerns, but he would be the one to take action and he would do things his way, not necessarily the best way. Inwardly I dreaded his decision to involve the programmers in the documentation because I knew they not only hated writing but also were inept at it. They could not grasp the user's perspective and didn't much care as long as they could sit in a dark room and code all day. I thought about a quotation one of the programmers had scribbled on the whiteboard in Hugh's cube: "You start coding. I'll go see what the customer wants."

My manager hit the Speaker button on the conference room phone and dialed extension 14. After three rings, Hugh's manager, Bryan, answered. Bryan was the director of development and was very protective of the programmers. He didn't want them to spend time on anything other than coding, and he discouraged people in other departments from consuming their time in any way. Time spent helping nontechnical people was a waste of time, not to mention a waste of money. The less time spent on coding, the less likely the programmers would meet the software release dates. They needed all the coding time they could get.

As my manager and Bryan discussed how the programmers should assist me with the documentation, I began to feel wistful and nauseated. I hadn't yet learned when to keep my mouth shut and when to voice my concerns. Receiving help from the programmers would be more time-consuming than putting in the extra hours and doing it all by myself. None of the programmers understood how the software product worked as an integrated whole. Programmers worked on specific features of, or enhancements to, the software product and concentrated only on the coding that concerned them. Additionally, they were isolated from the user. They never visited client sites and had never seen how the software was used on a daily basis. As if those dynamics were not sobering enough, none of the programmers were particularly fond of writing, and I knew only one of them to be capable of writing in a somewhat coherent, user-friendly manner. I wondered how I was going to edit all of their documentation within the next fifteen working days.

Satisfied that he had tackled the documentation issue successfully, my manager proceeded to the next person at the table, then the next, making the rounds and feeling important while the rest of us felt disillusioned and bothered. I drowned out my peers and their problems as I concentrated on my own. I silently contemplated asking my manager to forget about involving the programmers in the documentation, but I could already hear his reply:

> *Documentation is a formality. Users don't read the documentation; they call the technical support line. The programmers can help you document the features, and we'll have documentation to ship with the product. Nobody will read it, but we'll look good simply by making it available. Something is better than nothing.*

> *The users don't read the documentation,* I refuted him in my head, *because it's not intuitive. They don't know where to find the information. And if by some chance they do find it and it's not helpful or clear, then of course they're not going to read it. If the programmers write the documentation, the users will not read it.*

It was grand to argue with the director of marketing in my head because I could control the outcome of our conversation. He knew nothing about documentation, and I resented him for making big decisions regarding users whom he didn't know or care about. He cared about marketing, not about users. I decided secretly, in that room full of angry, stressed-out people, that I would devote myself to creating clear, concise documentation for the user by myself. I would accept documentation submissions from the programmers but would not use them. The programmers would never know the difference, and my manager would never review the documentation, so I resigned myself to the working eighteen-hour days to do the job right. Meanwhile, I was stuck in this stupid meeting.

The meeting eventually ended, and everyone filtered out of the conference room. I trudged back to my cube, tossed the blank notebook and pen on top of a stack of unresolved software bug reports, sat down, and began typing. I ignored the warm throbbing pain in my right wrist and concentrated on the rhythmic tapping of my fingers on the keyboard. This keyboard was going to be home for me for quite a while.

Elna R. Tymes

Diary of a Tech Writer

Author or coauthor of thirty-three books on software, Elma R. Tynes is president of Los Trancos Systems, a Silicon Valley communications and software development company. She is also a longtime observer of life in the computer industry.

The process described below is, of course, fictional, as is the client company and all personnel described. However, the incidents described have actually happened to the writer, at one point or another, in the course of producing technical documentation.

A potential client called today to set up an interview. They have a project that needs some help, and based on my résumé, they think I might be able to get the job done. I'll have to update my résumé and find some appropriate samples to bring to the interview. The client is a department within a software company, and the project will document an administrative tool. Sounds simple enough.

The Interview

I think the interview went well. I had some samples from other documents I've written for database applications, and we talked about the documentation process. Then the documentation manager explained the application and how it wasn't quite done yet. I asked about the target audience (it's network administrators) and the due date (two months from now), and we talked about the platform on which the documentation needed to be developed. It turns out that, although the eventual destination is UNIX machines, all the draft work can be done on PCs. I can even do a lot of the writing on my PC at home, although I'll have to come in to use the application on the client's network. She wants the document in FrameMaker, and she wants a Help file to go along with it.

So here we are with two weeks to go, and I still don't know

1. *when the final software will be available for me to compare with what I've written,*
2. *when I'll see review comments on the Help file,*
3. *or whether my FrameMaker files will accurately translate into HTML files.*

And there are fresh, young college students out there who think they want to get into this business!

The Proposal

It's a good thing I've written proposals for projects like this before. At least I have a template to follow. After the interview the project manager gave me a copy of a document similar to the one I'd be producing, and we discussed what would be similar and what would differ, so it was easy for me to develop a reasonable document outline for the proposal. I remembered to include sections on having the client provide access to adequate source information, the schedule and what the various milestones mean (and what happens if they're not met), what exactly I'm expected to deliver by the end of the project, my assumptions regarding tools and stability of the software and what could happen if my assumptions weren't met, and how I'll bill them.

Previous experience has taught me the importance of stating my legal relationship to the client company. (I'm technically employed by my own company, so the client doesn't have to deal with W-2 issues; however, their human relations department insists that I furnish proof that I have the necessary workmen's compensation insurance, legal documents proving that I'm allowed to do business in this state and city, and an insurance rider on our company insurance policy that covers any other liability issues.) So the proposal spells out that I'm an employee of my company and that the proposal is a contract for a company-to-company relationship. It also specifically states that we expect to be paid on a basis of net thirty days.

I sent the project manager two copies of the proposal. She signed it but had to send it up her management chain for further approvals. She hoped I could at least come in and meet with the developers while waiting for the approvals. She sounded so desperate that I figured I could at least do that.

Meeting the Developers

There are five developers on the team, each responsible for a different part of the application. Their project manager has done several other projects, but none involving a Help file. He asked me to explain what the developers needed to do with a compiled Help file—how to hook it into the application and make everything work. Only one of the developers had ever tried to attach a Help file to an application, but he seemed to know what he was doing. The team agreed to let him handle the final effort in attaching the Help file.

While they were explaining how the application worked, there was some discussion about how the interface would look. As I listened, I realized that they hadn't yet agreed on what would be included in the administration program and that it was not even to the prototype stage. I asked when they intended to freeze the interface; they all looked somewhat stricken and glanced in the direction of their boss. Uh oh! I pointed out that what I wrote would mostly reflect what was on the interface, and he decided that they'd have to put more effort on that in the near future. I gently pointed out that they were now slightly over seven weeks from the delivery date, with the clock ticking. He looked at the team, and eventually they agreed that they could have a prototype ready for me to see in about a week.

I reported the results of the meeting to my project manager, who promised she'd talk with the development manager about getting the interface frozen. She also told me that the finance department wouldn't agree to net thirty day terms in our contract—the company policy is net forty-five days. That bothered me—I've already started working on the project, and I won't see any payment for my work until the project is nearly over. I could stop work until we reach a better solution, which wouldn't earn me any friendships, or I could simply get on with things and pray that I don't have any financial misfortunes until the payments start coming in. I learn a lesson from this: get a payment up front, sort of a down payment, from this company before starting work on any future contracts.

The writing can be done on my own PC, but the research is going to have to be done on site. The company's firewall and security system won't allow me remote access to the system I'm documenting, so I'm going to have to come in daily for a period until I really get an understanding of what I'm documenting. However, once I start writing, I can e-mail my document files to the documentation manager, and I can reach the development team by e-mail if I have any questions. They can't set up an e-mail account for me here, though, because it will take too long to get the necessary approvals. That means taking screen shots on site, saving them to a zip disk or something portable of equivalent storage capacity, and inspecting them while I have the document file open on my machine at home.

Initial Research

My documentation manager got the necessary signatures on Tuesday of the second week, so I felt I could start the research the next day. That means we've wasted one and a half weeks of an eight-week project with meetings and proposals and getting approvals. My manager is now expressing concern that the project will not be finished on time, but I've assured her that I've done projects like this before and met the deadlines. However, her concerns are now yet another thing I have to manage in this project—it isn't enough to do the work, I also have to make sure she sees enough progress on a regular basis that she feels secure that I'll finish on time. This kind of project can really use a weekly status report, so I'll make sure she gets one every Monday morning along with my weekly time sheet.

It's now Friday of the second week, and I've been using the system and reading everything I can get my hands on. The good news is that somebody wrote a functional specification, outlining how the product and its interface are supposed to work; the bad news is that the interface is still not ready for me to see. The developer promised to have it ready for me next week.

Because of the functional specification, however, I've been able to begin writing pieces of various chapters. My manager has done this before, so she understands why I won't have complete chapters for her on Monday. However, she expects several completed chapters by the end of next week. I think I can do that—some of them explain syntax and formulas and search rules in a general way.

Today it's Wednesday of the third week, and the developer in charge of the interface was able to show me what he'd been working on. He asked my opinion about dialog boxes versus using a fill-in-the-blank approach, and I told him how I preferred to work. I also pointed out that, for some users, moving a hand back and forth between the keyboard and the mouse was an irritant and that it was preferable to complete a set of actions either with the keyboard or with the mouse. He appreciated my input.

So today I've been working with the new interface, writing down everything that I do so that I can write the chapters from the user's perspective. I've stumbled across a few things that didn't work and pointed these out to the developer, who was glad to have me do some initial quality assurance work. My manager told me where to find a copy of a graphics program so that I could take the necessary screen shots; I've stored all of those screen shots on zip disks and on an archive file on site so that I won't have to schlep the zip drive back and forth all the time.

First Draft

It's now Monday of the fourth week. The document outlined in the proposal consists of twelve chapters and an appendix. I've completed nine of the chapters and the appendix and am reasonably certain I'll have the other three chapters done by the end of this week, as promised. However, the development manager e-mailed me this morning, indicating that there would be changes to the software, probably this week, because of a new file format the product would have to interpret. I'm not sure just what impact those changes will have on my chapters, but I'm trying to put a reasonable light on it in the status report to my manager. The development manager said he'd know what would need changing by Wednesday morning.

My manager also said that, in addition to the FrameMaker files, she'd want the final files to be in Postscript and .pdf formats, so that they can be read with Adobe Acrobat. That's no problem—FrameMaker lets you save files in different formats once you're through changing the contents. However, she is also getting pressure from marketing to translate the document files into HTML and post them on a customers-only part of their intranet. That could be a problem, given that FrameMaker's styles don't always translate smoothly into HTML. I'll have to research that one and experiment with one or two of the short files.

Today is Thursday, and I have to deliver all the chapters and the appendix tomorrow, as first drafts. I'll need to compile them into a book and run the table of contents, but I can leave the index till later. I just realized I need to include a glossary, too. But all the chapters but one are done, the appendix is done, and the front matter is done. It's going to be a long night tonight, but I'll have that last chapter ready tomorrow, too.

I did it! Today's Friday, and I delivered the entire book to the developers for review. They now have a week to read it all and get any corrections or changes back to me. My manager is very, very pleased—I've made her look really good to her superiors. And while they're reviewing the book, I'll outline the Help system and see what I can use from the document files.

First Review

My manager said she read the entire book and is happy with the way I've explained things. She pointed out a few stylistic differences between the way I said something and the way their marketing and legal folks prefer to say things, and she suggested I make some minor changes to a couple of formats in the template I've been using. So far, so good. Let's hope I get some good feedback from the developers.

It's now Thursday and I haven't seen anything from any of the developers. I asked their boss about feedback yesterday, and he said that, as far as he knew, they were reading through the chapters. I'm going to casually chat with a couple of the developers and see if they've spotted anything seriously wrong.

Wouldn't you know it? It's Friday morning and the developers haven't even opened their drafts. I couldn't find the guys I went looking for yesterday—I know they have had debugging problems, but I assumed they were reviewing the manual, too. I'm going to have to ask their boss to insist that they review the manual.

It's Monday of the fifth week, and still no feedback from the developers. My boss knows I'm concerned and offered to put some pressure on the developers' manager. I can't think of anything short of bribery that will get them to review the manual. If I haven't seen anything by late this afternoon, I'll send them a note promising freshly baked chocolate chip cookies for anyone who turns in their review copies tomorrow.

It worked! I brought in three dozen chocolate chip cookies today, and three of the developers brought me their marked-up review copies. The other two promised that they'd get me comments tomorrow, if I'd just let them have two cookies each. Their manager gets any that the team doesn't get, so there's some good-natured bantering going on.

In looking over the review comments, I've noticed that two of the developers have markedly different ideas about how the product works. I'm going to have to spend some time this week with both of them, trying to straighten out just what to say. Other than that, there are some really useful comments and corrections, and I'm incorporating them into my files.

Second Draft

The second draft, incorporating all the comments and corrections, has gone out on schedule, at the end of the fifth week. I hadn't planned on baking chocolate chip cookies as part of my technical writing efforts, but if that's what it takes to get useful feedback, that's what I'll do. I've also sent along the Help file, compiled in RoboHelp, and asked them to comment on it.

Second Review

We're now into the sixth week of the project, pretty much on schedule. The developers have until Wednesday of this week to get comments back to me and have teased me about what form the bribe would take this time. I'll take them all out for Chinese food if they get the comments back in time—I can always write it off on my taxes as a business expense.

I *knew* things were going too smoothly! Today is Thursday, and the developers' manager said that a few early beta users of the program have requested changes to the GUI. That means I'll have to reshoot—and re-explain—every screen that is changed. The manager promised me that he would personally review every changed chapter to make sure that all the corrections were made, but I suspect I'm going to have to change about 20 percent of the screen shots and reread most of the book to make sure that any cross-referenced material reflects the new GUI.

And my manager says she now wants the final chapters translated from FrameMaker to HTML on a UNIX platform and that there are formatting problems in taking the PC files into a UNIX environment. I'm going to be here all weekend, I can tell, and the only good part about that is that I'm being paid by the hour.

Two of the developers returned their review copies to me on Friday, but said they hadn't had time to look at the Help file. The other three are apparently too busy with the software changes. I've mentioned this to the development manager, who said he'd appoint someone to review the Help file for me. However, he also said that he and his team haven't decided on just which screens will need to be changed.

So here we are with two weeks to go, and I still don't know

1. when the final software will be available for me to compare with what I've written,
2. when I'll see review comments on the Help file,
3. or whether my FrameMaker files will accurately translate into HTML files.

And there are fresh, young college students out there who think they want to get into this business!

It's now Wednesday of the seventh week, and I just received a master list of the new screens and how they differ from the previous screens. The developers will be burning the midnight oil the rest of this week getting everything done—they're supposed to have the product ready for formal testing by Monday of next week.

One thing that's become obvious is that the flow chart in one of the chapters will have to be redrawn. The changes to the screens reflect a few major changes in logic, and since the developers have been using my flow chart as a means of organizing the changes, I'll have to get one of them to explain what they've changed. Redrawing the flow chart isn't all that difficult, but I did it using my copy of Visio at home since the client doesn't have Visio on site. More schlepping of the zip drive, I can tell.

It's now Monday of the last week. The developers were here all weekend, working on the new screens. They've just given me the marked-up flow chart, and the logic changes are relatively minor. All but two of the screens are now done, and they've said they'll have the remaining two done by the end of the day. They managed to slip testing by a day, but they have to produce a clean, compiled version by tomorrow morning. I'm staying out of their way today—I'm taking new screen shots and trying to identify where I need to change the text.

Final Draft and Review

It was a long night for me, too, but it's now Tuesday, and I've managed to incorporate all the new screen shots and change the related text. Two of the developers have promised to review the manual one more time, and since I'm having to change the Help file, the developer assigned to that file has promised he'll look it over on Wednesday.

However, we have yet another problem: the lone developer who knows how to attach a compiled Help file to the application has come down with the flu. None of the other developers know how to do this. The program has to ship with the Help file attached, and that's supposed to happen by the end of day Friday. The developer's boss said he'd look around and try to find another programmer who knows how to work with Help files.

It's a sunny, clear Thursday morning, and the developers just put their review comments about the manual on my desk. And the developer who said he'd review the Help file said he'd have his comments for me later on this morning. We may make the deadline yet.

Oh, yeah. In my mail last night was the first check for payment on this job. Forty-five day payables. The things we do for money.

Production

I got all the review comments incorporated last night, both into the FrameMaker files and into the RoboHelp files. Everything was due today, and I started the morning hoping that things would go smoothly.

It was easy to save the FrameMaker files as .ps and .pdf files, but saving them as HTML files was more difficult. There were still the formatting problems in the HTML version, and a programmer had to tinker with them to get clean HTML files. However, my manager got her HTML files by 4:00 P.M.

And the developers' manager did find someone who knew how to hook the RoboHelp files into the application, and the problem was fixed by 3:30. So a little after 4:00 P.M. my manager and the developers' manager went trooping off to the release department, having uploaded everything into the waiting file system, to announce that they had, indeed, delivered as promised. They came back about fifteen minutes later and announced that they were taking the entire team out for beer and pizza.

While we were celebrating our on-time delivery, my manager asked if I'd be willing to stick around for an extra two weeks. It turns out that marketing had requested even more changes to the application, and the developers' manager had negotiated a "minor upgrade" to be released in two weeks. *Project completion* is a very relative term in this business!

Steven Jong

Samurai Review

Steven Jong, a second-generation technical writer, has over twenty years of experience in the field. He holds a bachelor's degree in physics and astronomy and a master's degree in science communication from Boston University. He has taught classes in structured documentation quality in the United States and Europe. Today Steve is a documentation manager and a senior member of the Boston Chapter of the Society for Technical Communication; he is active in the chapter's publications competition. He also writes about documentation quality for the STC Quality special-interest group. Steve lives in Westford, Massachusetts, with his wife and three children.

Paul Trask, the test engineer, sat at the workstation in his office and started up a copy of the complete Pharmaceutical Assistant program for the first time. He studied the cheat sheet the engineer had e-mailed him, dug into the pages of the massive *CRC Handbook of Tables for Organic Compound Identification* in search of something simple to make, and on impulse chose dimethylethylamine. The program prompted him to enter the compound's chemical parameters: molecular composition, chemical bonds and their angles, handedness. When he finished, the program began to cogitate, spitting out a series of in-progress messages that let Paul know it was making headway. After about a minute it output a stream of chemical reactions embedded in Lisp constructs. He compared the results with the description in the reference book and nodded. The program had come up with a plausible formulation. Paul was impressed; he had just watched a computer program do the work of a professional chemist.

He turned to the description of another molecule, synthesized from the same compounds and not noticeably more complex than dimethylethylamine, and began typing. The program thought about how to synthesize the compound and went into a tight loop:

> *Dave looked at his mutilated draft and sighed. "I think it was H. G. Wells who said there is no impulse so strong as the urge to edit someone else's writing. I feel as if John Belushi should come in with a samurai sword and slice my book in half."*

Lookup . . . Chaining . . . Lookup . . . Chaining . . .
Lookup . . . Chaining . . . Lookup . . . Chaining . . .
Lookup . . . Chaining . . . Lookup . . . Chaining . . .
Lookup . . . Chaining . . . Lookup . . . Chaining . . .
Lookup . . . Chaining . . . Lookup . . . Chaining . . .
Lookup . . . Chaining . . . Lookup . . . Chaining . . .

Paul checked the computer's activity on his other screen; the CPU was pinned to the wall, stuck running the same portion of the program endlessly. He interrupted the program to put the system out of its misery and reached up to pencil an *X* on a tally sheet pinned to the wall above his desk. By now there were 150 *X*s on the sheet, each representing one STAR he'd written against the Pharmaceutical Assistant.

As he restarted the program, Dave Williams, the contract technical writer, came into the office, wordlessly leaned over Paul's shoulder, and stared at the screen. "This is the latest and greatest," Paul told him.

"Son of a gun. I've never seen it running." Dave tried to match what Paul was doing with the descriptions he'd written.

"I just got it this morning."

Dave tapped his watch. "Well, are you coming?"

"What?" Paul looked up. Dave was carrying two green pens, a pad of lined paper, and a copy of his first draft nestled in his arm as if it were a baby. It was nearly twice as thick as the book in Paul's lap.

"Your review meeting? I wasn't planning to attend. I already gave you my comments."

Apparently this wasn't what Dave wanted to hear. "Yes, but I was hoping to have at least one friendly face there."

Paul observed, "The book looks bigger than I remember it. Have you been powertooling?"

"Yeah, I added a lot of pages in the last week." That was an understatement; Dave was actually giddy from sleeplessness.

"If it's changed that much, maybe I'd better go." Paul scowled at his watch. "I suppose I could spare an hour."

"Thanks. That's great. I really appreciate it." They walked over to the conference room together.

"To tell you the truth, I enjoyed reading it. I can really get a sense of what PA is going to do."

"That's nice of you to say. It's been tough with so little to go on."

"Tell me about it. I've had to test bits and pieces. It must be frustrating for you."

"Well, I make the straw man, and the reviewers set the torch to it. But it's my job. Besides, at this point I've done most of the work. It's all downhill from here."

"Wish I could say the same."

"What's the protocol for review meetings at this company?"

"They're usually pretty boring. We show up, you walk through the manual, we give you comments, you get your questions answered, and everyone leaves. It shouldn't take more than an hour or two."

"Even for this?" He heaved the draft up a little.

"Well, maybe longer for that. I've never seen a manual that big from this company."

Dave took a certain grim pleasure from that remark. Whenever he finished a draft, he always liked to think that it was without error, though he knew it was never true. But whatever its flaws, it was his baby.

They reached the conference room, and Paul ushered him inside. At the far end of the table Bob Beauregard, the lead engineer, sat with two other engineers, each with a copy of the manual stacked in front of them, looking like a high-tech tribunal. There's only three of them, Dave thought. This should be a piece of cake.

Paul knew Bob and one of the others, a huge, sallow-faced man-child with a mangy ponytail. A large black calculator and an ostentatious set of keys hung from his belt, which was already strained by the imposing gut stretching his flannel shirt out like a spinnaker. As they entered he was saying to Bob, "So I told him it was just a simple matter of programming!" and while the other two merely simpered, his woodpecker laugh echoed in the windowless room. But when he saw Paul, his smile degenerated into a sneer.

"Look, guys, it's Paul Crash. Are you still filing false STARs?"

"No, George," Paul said wearily. "Are you still corrupting your pointers?"

Dave walked to the group and laid down his things, but Bob shook his head. "Not here." He pointed to the far end of the table. "You can sit down at that end and take notes."

Dave hesitated, smiling foolishly. "What? Are you kidding?"

"No, I am not kidding," Bob smiled back. "Devos only at this end. Doco on that end. Paul, you are invited to join us."

"I'll sit down here, thanks," he replied. He sat with Dave at the far end of the table.

Dave uncapped his pen and said, "I don't think I know everyone here. I'm Dave Williams." The engineers did not respond. What a bunch of stiffs, he thought. Paul said quietly, "You know Bob, and that's George Macon, but I don't know the other one."

Dave looked up. "I'm sorry, what was your name again?"

"My name? Is Stephen Epstein." He had large brown eyes and a thin moustache. His hair was a nest of cowlicks rising in all directions. He wore a blue long-sleeved dress shirt buttoned all the way to the throat. Looking at his preternaturally colorless skin, Dave wondered if they'd hired him or fished him out of the river.

Bob said, "Steve's our brand new gooey expert. He's sitting in today to get up to speed on the process."

"Oh, is this your first review meeting?" Dave asked conversationally.

The engineer looked at him blankly. His jaw fluttered, as if Dave had asked him to summarize Proust. Finally he responded, "Ah, yes—no—ah, in a, ah, in a manner of, ah—" He waved his hand impotently. "This is not my first, ah, professional position, and I have participated in—ah, in attendance at reviews—documentation reviews, of course, in other, ah, instantiations."

"Oh." My God, where did they find this guy—Mars?

Bob brushed potato-chip crumbs from his hands onto the carpet. "All right. This is how I want to do it. Rather than kill a forest to print copies of the manual, I split up a master copy and circulated the pieces." He touched the fat three-ring binder next to him. "George and I have comments of our own. If you have any questions, I'm the arbiter, since it's my ass on the line."

Dave thought the meeting was about his own ass, not Bob's. He'd never encountered the practice; perhaps it was a Beauregard brainstorm. He wanted to assert himself before he completely lost control of the meeting, so he smiled and said, "Well, of course I look to you to resolve any technical disputes, but I'd like to reserve judgment on any stylistic comments."

"No," Bob said, "I'll tell you what's a technical comment. You can wordsmith the rest later."

Paul, as a dispassionate observer, thought only, Isn't that what he just said? But Dave took it as a turf challenge and thought angrily, What's his problem? He considered his options. He could throw a shit fit, but contractors have no leverage. It had been hard enough to get information out of Bob without antagonizing him. No, his only option was to take it and like it. He replied equably, "Whatever. Let's see what you've got there."

Bob swilled coffee and turned to the first page. "I have comments on the preface." They all opened their copies to the preface page, where Dave had written:

> *The AIGen Pharmaceutical Assistant (PA) is a sophisticated application designed to assist organic chemists by simplifying the process of pharmacological synthesis.*

"It's an AI application, you've got to say that."

"All right." Dave marked the change neatly in green ink in his copy.

Macon said, "I don't like the word 'sophisticated.' Is that what we want it to say? It might have legal implications. What does 'sophisticated' mean in this context?"

"I meant to say it's powerful and complex," Dave told him.

"Then why not just say 'powerful and complex'?"

"Don't you think you might scare people off with a word like 'complex'?"

Macon shot back, "Do you think the work we've done is simple?"

"No. I just don't want to say it's complex."

"I think you should describe it accurately, and it is accurately described as 'powerful and complex.' In fact, it's *very* powerful and complex. People have a right to know what they're paying for."

Dave frowned. "Well, if we're getting into adverbs, that's a matter of style, not technical accuracy. I'll think about your suggestion."

Macon said loudly, "I am not giving you a suggestion. I am telling you how to accurately describe the product."

Dave looked at him impassively. Maybe they were playing good-cop, bad-cop with him. He pursed his lips and said, " 'Very powerful and complex'?"

"That's right."

He bent to write it down. "It's your product. If you want it to read that way, fine."

Bob said, "Add 'state-of-the-art'."

Macon asked, "How do we abbreviate that? SOA?"

Dave said, "You don't abbreviate that."

Bob nodded. "I think we abbreviate that."

"Hey, that would be funny," Macon said, brightening.

"No."

Bob joked, "That's our acronym du jour."

Dave tried again. "Are you planning on using the term again in the book, or just once in the preface? If you're only using the phrase once, you don't need to turn it into an acronym. Then readers have to absorb two things."

Macon said, "Just in the preface. I don't know. I didn't write it."

"Well, I did, and I say it doesn't get abbreviated," Dave said as firmly as he could. The engineers looked at each other and shrugged. Bob said, "Every dog has its day."

Macon went on, "I don't think 'assist' has the right flavor. It implies action. Change it to say something like it 'provides assistance.' "

"That weakens the sentence," Dave pointed out.

"Why are you so defensive? This is the most important sentence in the manual. We should make it say exactly what we want it to say."

Bob said, "Why are you limiting the audience to organic chemists?"

Dave asked, "Who else would read this?"

"Research chemists. These heuristics can be applied to chemistry in general."

Macon said, "The engineering staff will want to read this to see how we did it. They'll get some pretty good tips."

Bob frowned. "What do they call themselves? Are they really organic chemists?"

"You work with them. Don't you know?"

"Jay-sus, I picked their brains, I didn't read their nameplates. I think 'organic chemist' is okay. But you've got to add something like 'research chemists and programming staff and other engineers who wish to learn how this application was designed—' "

"—'and coded'—"

"—'and coded,' great, you got that?—'as well as . . . ' Can you think of anyone else?"

"Just say 'et cetera.' That should cover it."

Epstein spoke up. " 'Simplifies' is too, ah, too, ah, overly—I don't think you can call it simplification, when more accurately it would be more at 'provides useful advice.' "

Macon said to Dave, "How does PA do this? You don't say."

"This is only the first sentence," Dave replied, feeling dizzy. "There's nine hundred more pages here. You don't have to cram everything into the first sentence."

"I think it should say something about how the program works."

Bob said, "It advises on the process of synthesis of pharmacologically active substances."

Dave was frowning as he tried to keep up. "This sentence really doesn't sound too good."

Bob said enthusiastically, "I've got it. 'It provides assistance'—shit, I don't know what it says anymore. Read it back to us."

Dave scowled and tried to pick through the layers of changes:

> *The AIGen Pharmaceutical Assistant (PA) is a very powerful and complex state-of-the-art artificial-intelligence (AI) application designed to provide assistance to organic chemists and research chemists by providing useful advice on the process of synthesis of pharmacologically active substances. It is also of interest to programming staff and other engineers who wish to learn how this application was designed and coded, etc.*

Paul said, "I don't know what it means any more."

"It sounds like an insurance policy," Dave groused.

Macon said, "It sounds just fine to me. It's precise and accurate. You could learn a thing or two about writing if you accepted constructive criticism instead of acting like you're the only one with the right answers." He turned to Bob and said, "I told you this is easy. He's stealing money doing this."

"I need some leeway in how I phrase things," Dave said. "It has to read well."

Macon waved dismissively. "You can go off and play with the commas all you want. Just make sure it says all these things."

Bob shook his head. "No, he can't, not when the technical accuracy of the book is at stake. I don't want any changes to what we've agreed to. Now let's move on. We don't have all day."

Paul fidgeted and looked at the wall clock. They had spent forty-five minutes wrangling over the first sentence. He wished he was somewhere else. Dave's taking a terrible lot of abuse, he thought. How can he stand it? I'd have lost control by now. In the domain of the English language, the engineers were obviously deaf to nuance.

Bob said, "In the recommended reading list, you need to add more titles. We put together a list for you." He slid it across the table.

Dave picked it up and flipped through the pages with mounting discomfort. "There must be fifty references here! I think that's way too many."

"Which ones don't you think are necessary?"

"I don't know. But you can't expect users to read fifty books before they start to use PA."

"Why not? They have to understand how PA works before making use of it. These are not software engineers. All they know is organic chemistry. They have to be familiar with basic computer science concepts."

"Database," Macon said.

"Exactly," Bob agreed. "And fuzzy logic. And predicate calculus."

Dave shook his head doubtfully. "Do you have time to read fifty references?"

"No, I don't have to," Bob said. "I already know them."

Epstein cleared his throat. "Do you, ah, are you suggesting—ah, given the background of our user, ah, environment, or community—our mainstream user community, that, ah, since they are primarily chemists and not software engineers, ah, that, ah, that they ought to see—at least some of the material Bob references—are you saying you are willing—ah, prefer to incorporate the material he references in this document?"

"What?" When Dave untangled the question he was horrified. "Oh, no. I see no need to add to the manual. It's big enough as it is."

Paul chimed in, "If we add any more pages, no one will be able to lift it."

Bob thought about that for a moment. "We'll compromise. Put the list in for now and I'll think about maybe paring it down or incorporating some of the material later."

Dave tucked the list into his copy. "Are we done with the preface now?"

Macon turned the page on his copy. "Now we get to the main problem I have with your manual. Your first illustration—the one that shows what the screen looks like—it's wrong. We're changing the user interface all around."

"Yeah, I just saw it in Paul's office. I'll take care of it."

"That's just a prototype. Steve's redesigning the whole thing."

"What?" Paul sat up. "How much is it going to change?"

Bob spread his arms theatrically. "Who knows? I thought it was fine the way I designed it, but some suit decided to put his thumb on the scale, and what am I going to do? I'm only the project leader. Steve, why don't you tell them what you're doing?" Dave thought he smiled paternally; to Paul he seemed to smirk.

Epstein ran his hand through his hair compulsively as he spoke. "There will be substantial, though I would not say an entire, ah, change, but, ah, certainly significant changes to the, ah, user-visible aspects of the interface, in, ah, in the area of graphical molecular display, which I have taken to calling, ah, for want of a better phrase, perhaps you could suggest one, ah, I have been calling it, ah, taken to calling it the 'Tinkertoy look.' I'm afraid, ah, I regret to say, and in retrospect I'm sorry that I neglected to bring screen captures, so I cannot show you—perhaps it would be easier, ah, clearer, if I could draw something . . . "

He jumped up to the whiteboard and rapidly began to sketch out what he was working on. It looked to Dave as if he was changing the way the program displayed molecules from Bob's eccentric molecular formulae to a more realistic and customary design of colored balls stuck together that he could display pictorially. Designers would be able to describe the molecules they wanted to synthesize by pointing to a palette in the shape of a color-coded periodic table of the elements and dragging out colored balls representing atoms. They would then marshal the atoms around the screen to set bond lengths and angles. It was a wonderful concept that any chemist would grasp immediately.

Epstein drew molecules and formulae like some mad graduate assistant. He got gaseous the longer he spoke; he belched, stuttered, snorted, giggled, and sighed; rubbed his nose and unconsciously stuck his finger in. He rarely turned to look at his audience, so most of the time he spoke right into the whiteboard. As hard as he was to listen to, he was worse at expressing himself visually, because he drew with his right hand and erased with his left, blocking with his body what he was describing until it was gone. Dave scribbled notes furiously, straining to hear and craning his neck back and forth to see around him; finally he had to get up and stand right next to him. He wished he'd brought a tape recorder, or better yet a video camera. Not a jot of anything Epstein said had ever been recorded in a specification or design memo, so everything came as an unpleasant surprise; and his experience screamed out that none of it ever would be written down, so this was his only chance to capture it.

Bob watched intently, from time to time interrupting to interrogate or raise objections: "You'll need twice the CPU horsepower to do this. Will the size of the balls be proportional to the size of the orbitals? How can you represent double and triple bonds?" Epstein cogitated over the challenges and addressed each one in halting detail. As he talked, he invoked more and more startling images, from pop-up windows for entering data precisely, to named molecular substructures, to rotating the molecular view in three dimensions. When, after twenty minutes, he sketched an animated moving point of view that would allow users to inspect the molecule from all perspectives as if they were flying through it, Dave, by now thoroughly terrified, had to ask, "Are you doing all this in the first release?"

Epstein shook his head until his hair wiggled. "No, ah, I'm sorry if I gave you that, if I inadvertently suggested that these were all committed enhancements, because I was only discussing, ah, possible futures, suggesting possible future, ah, enhancements for discussion."

"You started by talking about things you're doing for this release, and now you're talking about things you'd like to do in the future?"

"Ah, yes, ah, yes," he nodded, his chin vanishing completely.

Dave sighed. "When did you stop describing reality?"

Epstein's jaw worked, and he raked his hand through his hair, but he made no answer. Bob said jovially, "Hell, I don't think he even knows! He'll do as much as he can."

"When am I going to get my hands on something?" Paul asked.

Bob said, "Not for at least two weeks."

Dave sagged as the realization sank in of how much more work lay ahead of him and how undefined it was. "I'm going to have to redo every example in the book."

Macon said suddenly, "That's what I'm here to talk about. Putting in all those examples before we've finalized the interface is inherently wasteful. I think you should just describe what PA does without showing any specific user interface. That would be a much more effective presentation." He looked pleased with himself, as if he had independently stumbled upon a master stroke of technical documentation. "All you need to do is devise an abstract notation, like Backus-Naur Form. It would be independent of any user interface. Imagine how efficient that would be for you! We could make changes without affecting the book."

"You mean I'd write something like 'Specify a chain of carbon atoms' without telling them how to do it?"

"Exactly."

"But how will they know what to do?"

"Once they know the convention, it won't matter. Besides, you've seen what he's doing. It's genius! It'll be intuitively obvious; don't worry about it."

"How could they be useful examples without showing what the results would look like? Users would have to learn two conventions—the interface and the way I was describing it. That would be much too abstract to follow."

"It doesn't sound difficult to me," Macon said sharply. "I work with levels of abstraction every day. We could complete the user interface without having to worry about you slowing us down."

Bob nudged Macon's arm. "I think he means it would be too hard for him to write." He put his feet up on the table, enjoying the show.

Dave gritted his teeth. "Should I document what you've described here, or should I wait for it to settle down?"

"No, this is cast in concrete. We don't have the time to change it."

"All right, I'll try to figure something out. Now can we just get on with it?" He nodded wearily at the three-ring binders. "I think you've got a lot more comments in there."

"Now that you understand the changes coming up, we can go through these quickly," Bob said agreeably.

He slid across the communal review draft. Dave leafed through it in horror. Apparently Bob had turned over the task of reviewing the book to a team of New York City subway artists whose favorite color was blood red, he thought. The pages were so heavily worked that they appeared swollen, stained with coffee and what looked like water, which, considering how engineers worked, probably meant bathroom splashes. It seemed as if every page was covered with comments that crammed the margins and spilled out to spoil his neat typography with crude scribbles. It would require word-by-word study to decipher.

Paul asked, "Can I see?"

Dave disgustedly pushed the copy at him. Almost every page was covered with changes. When he got over the initial shock, he looked more closely at what the engineers had written. One had penned "No" over and over, giving no indication what the correct information was. Another repeatedly opined, "Needs rewrite," which, come to think of it, was equally useless. A third favored a simple question mark, which Paul saw conveyed no information whatever. He recognized one notorious engineer's childish printed handwriting. He seemed to take delight in finding typos, but all his technical comments were enigmatic—"Not sure about this," "Must check," "Why is this necessary?" "I disagree," "Why would you want to do it this way?" and "Not always!" Many changes were in turn rewritten by someone else. One section was *X*-ed out with angry slashes of a red felt-tip pen; the only explanation was, "All that's missing is the smell."

Paul's ears began to burn. The comments read like bug reports he'd written in the middle of the night when his sense of restraint had weakened. He'd felt clever writing such remarks, but directed at his friend, they looked cruel and pejorative. Looking at Dave, who sat dejectedly next to him, he felt a sudden sting of remorse. He had always felt Macon's loathing was sour grapes because his STARs were right and Macon's code was wrong, but he began to understand how his reports could have been hurtful.

Bob tipped his chair back and looked up at the clock. "I'm starving. What do you say we send out for pizzas?"

Paul looked at his watch and said, "I've got to get going."

Bob said, "Are you sure you won't stay?" Beside him Dave scribbled a note.

"No, I've got a lot to do—" Dave slid the note to him. It read, DON'T LEAVE ME ALONE WITH THESE ASSHOLES!"—but I suppose I could at least have some pizza."

"That's more like it!" The engineers took a quick collection, and Paul and Dave pulled out a few dollars. "Hold on," Bob said, waving the wad at Dave. "We have enough. If you go buy them, we'll pay for your slices. Deal?"

Dave looked at him sourly. "No. Get your own pizza."

"Come on, don't be unreliable." Bob's eyes twinkled.

"Why don't we bring back something from the cafeteria?" Paul suggested quickly. "We could work and eat."

"That sounds like a great idea," Bob said.

The afternoon was as painful as the morning. The hours dragged; the engineers crawled through the programming examples in search of missing parentheses. Macon quipped that Lisp, the name of the language the Pharmaceutical Assistant was written in, stood for "lots of irritating superfluous parentheses," and the engineers laughed crazily at the joke.

When they got to his fifty-page appendix of error messages, Dave pleaded, "Is there any way to reduce the size of this list? Are users really going to see all these messages? Aren't some of them just for your own internal debugging?"

Bob shook his head. "No, they're all needed."

Paul said, "Bob, some of these are intermodule errors. I can't imagine users ever seeing them, and if they did, there'd be nothing they could do about them."

"At least," Dave said, "they should be explained a little. I had nothing to go on but the list. If someone looks up an error in here, all they'll see is 'there it is.' If we have to document them, we might as well say something useful. Can you provide explanations of what the messages mean?"

"No. I don't have the cycles to spare for it."

Dave felt completely shut out. "I think it's important," he pleaded.

Macon said, "We'll never get done with this meeting if you keep going down ratholes."

Paul said, "I've run into most of these errors. I think I know what triggers them. If you want, I can tell you which messages to keep, and when and why they occur. Is that all right, Bob?"

"As far as I'm concerned it's a don't-care. If you two want to go work that issue, fine. Now, Appendix B. We changed all these formulas, so you can dump what you've got here."

"Oh, God," Dave blurted. He had spent two nights until midnight formatting the seemingly endless number of chemical reactions PA knew about. Suddenly he felt exhausted. "I thought you wanted them in the book."

"No, those were stubs," Bob replied. "They were all dummy formulas of simple, well-known compounds. Didn't you think we were going to replace them with real data?"

Macon sneered, "You don't know much about organic chemistry, do you? These are trivial examples, just child's play. It's really only a proof-of-concept. What we're scaling up to is Beilstein."

"What's that?"

Paul said, "It's a German compendium of all known organic compounds. The standard reference." He turned to Bob. "You don't expect him to put all that in, do you?"

"Why?" Dave frowned. "How big is Beilstein?"

"Twenty-eight volumes," Paul said.

Dave's face went gray. "Then I think we should take these out," he said quietly.

Macon looked at his seventeen-button watch and said, "Shit, I have to flush the rest of this meeting. I'm late for another one."

"If you're already late, why don't you stay here?" Bob wheedled. "How could it be any more fun than this?"

Macon considered this, smirked, said "Maybe you're right," and settled back again.

"Are you really going to fit twenty-eight volumes of printed information into the knowledge base?" Dave asked. "How are you going to do that?"

"That's, ah, a good, ah, interesting, really a fascinating question," Epstein said. He launched into what sounded like a Ph.D. thesis defense on database design. Dave didn't understand a word, but he didn't want to admit it, so he nodded and suffered in silence. It was a full five minutes before Epstein petered out. Dave weighed the risks of a follow-up question, shook his head in resignation, and asked, "I'm sorry, but was that a yes or a no?"

"No, it was, ah, yes, that is, it's not as simple as that," Epstein said, warming to his subject. "What you need to understand is . . . " To the great amusement of the other two engineers, he launched into what turned out to be another five minutes of random brain dumping on why the database design was, as he put it in a flash of near-lucidity, infinitely bletcherous for this problem domain. All that Dave could gather was that the database of chemical formulas was going to be huge, that it would slow the program down markedly, and that he, Dave, was expected to write around the problem and make it sound like virtue. He had developed a fierce headache, and this late in the afternoon, Epstein's every word was like a ball-peen hammer against his skull.

In the end, listening to Epstein exhausted them all. Bob closed his book and rubbed the back of his neck. "Are there any more comments? Has everyone had a shot?" The others nodded. "All right. So when will you have these changes in? We go to the test sites in three weeks."

Dave rolled his eyes. "I'll do the best I can, but these are massive revisions. I'll have to take a chainsaw to the book. It'll look pretty shitty."

"Don't worry about it," Bob reassured him. "Nobody reads manuals anyway."

Macon said, "Remind me to bring in my Ph.D. thesis. It has more useful information than your manual, and it's a lot shorter besides."

Bob looked at his watch. "Jay-sus! Look at the time." The engineers stood up and walked out.

Paul went out after them. "Hey, Bob, can I ask you something?"

Beauregard stopped and leaned against the wall. "You've got about two minutes before my bladder ruptures."

"That draft wasn't wrong when he wrote it. You've added a lot of things you were calling 'bells and whistles' a couple of months ago."

"No, we were going to slap on a user interface eventually."

"That's not what you told us at the project meeting."

"What do you want from me? You're the one who wanted this cruft. That's what you get for writing so many STARs. I busted my ass putting fixes in. Isn't that goodness? Aren't you happy?"

"Why didn't you start with the user interface and work your way in? These changes are going to affect us, too. We have to requalify all your bug fixes."

"Because I've been working on more important things. Do you have any idea how hard it is just to name organic compounds? I've got it guessing right 95 percent of the time." He turned and walked away.

Paul went back into the conference room and found Dave calmly collecting his notes, as if nothing upsetting had happened. "That was a long day," he said.

Dave looked at his mutilated draft and sighed. "I think it was H. G. Wells who said there is no impulse so strong as the urge to edit someone else's writing. I feel as if John Belushi should come in with a samurai sword and slice my book in half."

"It was rough," Paul said sympathetically. "I didn't see that much to comment on. In fact, I thought it was pretty good."

"Thanks. You know, I've been in this business for fifteen years, and that's the worst review meeting I've ever been through. And it's not as if I screwed around on this. I worked diligently. I read the specs and documented what was there. No wonder I couldn't get Bob to answer any questions—he was too busy redesigning everything. You don't completely change the user interface three weeks before you show it to customers. You don't populate the database just before you go live with it."

"I agree the user interface changes are radical, but you saw what Steve wants to do with it—it'll be brilliant!"

"In five years, maybe. He may have a nervous breakdown by then. Did you see the way he was working his nose? I was thinking he kept his designs up there."

Paul shook his head. "I've seen worse. He'll be fine."

"Well, I'm pretty much going to have to start from scratch. No wonder manuals suck. I write it, they redesign it out from under me. It's like trying to change the tires on a moving car."

"Do you think you'll have it ready in time?"

"No way. I'll be lucky if I get half these comments incorporated in three weeks."

"What are you going to do?"

"Normally I'd ask my manager to push back on engineering to slip the schedule."

"Oh." Paul imagined their manager trying to push back on Bob's manager and drew a blank. "He's kind of hands-off."

"I knew that five minutes into my interview."

"If it's any consolation, I'm in trouble too. Most of the bugs I find are in the user interface. If they're changing it, I'll be flat-out for the next three weeks."

Dave shook his head. "Fortunately, I have an understanding wife." He picked up the copies.

Paul followed Dave all the way back to his office, where he removed the marked-up draft pages out of the binders and crammed them into his attaché case. "You know, if it's chilly tonight I think I'll make a fire and curl up in front of it with this book and a large glass of scotch. I'll take a sip, read a page, and throw it in the fire. See you tomorrow, Paul."

"Take it easy, okay?"

"I'm fine, thanks," he said quietly.

Paul wandered over to the window and saw him walking to his Civic in the twilight. Halfway there his attaché case sprang open and the pages tumbled out and blew around. He chased them like a man trying to herd pigeons and threw them back in. He forced the lid down with his knee and stuck the case under his arm, corners of pages sticking out everywhere. When he got to his car, he dropped his keys twice. Paul watched the little blue car dart onto the parkway and tracked it anxiously until it was out of sight.

Rochelle Gidonian

Thumbnails

Rochelle Gidonian was born in Kansas, grew up in Minnesota, and now lives in Israel with her husband and three children. She has worked as a secretary, nanny, cocktail waitress, high school teacher, script writer, science-fiction reader, and QA flunky. She earned a B.A. in linguistics and education from SUNY's Empire State College in Jerusalem. As electives, she took courses in the things that most interested her: computers and writing. Later, she took professional courses in computer programming and technical writing. She always enjoyed writing but wished that someone would tell her exactly what to write about and how, so that she could get on with the real work. Now she is living her dream as a technical writer.

I was at the top of my class during the technical writing course, having been both a professional writer and in QA, with some programming under my belt. Even before I finished the course, I was grabbed up by a smooth-talking subcontractor. Not a day went by that he didn't remind me how much I had learned while working for him, how much I owed him. Not a week went by without his throwing a temper tantrum. I think I could have forgiven him all of it if he hadn't stiffed me my last month's salary and benefits.

So it was while walking on clouds, overjoyed at my escape, that I changed gears and went in-house to work in the small technical publications department of a middle-size company. Everything was so different. I had a new desk and a fast computer. The intranet's practically unlimited memory was a godsend after the nightmare of the subcontractor's outdated, bugged, memory-depleted hardware. I had free access to the printer, the Internet, e-mail. I had my own phone and business cards. I had an almost-free membership at the gym. I had free lunches and holidays off. I had the option to telecommute every few days. My colleagues often smiled at my joy in such simple things—after all, a pleasant working environment is a basic necessity, isn't it?

One of the most pleasant aspects of this company didn't have anything to do with the latest equipment, the snazzy office, or the benefits. The best thing about working there was the absence of backstabbing—no gossiping, no jealousy, no loud political arguments in the cafeteria. Even when it came to the first real problem I faced, people were polite about their disagreements.

Suddenly it came to me: a solution. It was so simple, I slapped my forehead and said "Duh!" out loud. My solution was thumbnails. Excited, I explained my idea to the programmer. When I finished, she turned to another programmer in the office and said, "I like her! She actually listens to us."

The problem started this way: I came from a strong background creating online help and HTML-based help systems. The only other technical writer in the company (I said it was a small department!) had never done it. He had, however, taken a course or two in the principles of online help systems. The day he was assigning me the task of the online help system for the latest product, he said, "What is your opinion on how the help should look?"

I was somewhat at a loss. What exactly did he mean?

He continued. "Well, how long do you think each topic should be? I went to a course where I was told that each topic should be short enough to fit in the window without scrolling."

"I agree with the principle," I replied, "but if that means splitting a cohesive topic into two, I don't think that's right. I'll try to keep the topics short and to the point. I'll use the no-scrolling idea as a rule of thumb, but it's less of a bother to scroll down a few lines than it is to link to another page for no reason."

He considered this for a moment, then said, "Okay. I can live with that. So, what is your opinion of screen captures? Should they be in the online help system?"

"Absolutely not," I answered. Then I was taken by surprise as a look of triumph came over his face. What had I gotten myself into? I had an idea that I had just been called in as an objective expert witness.

That is exactly what had happened. The technical writer had had a long disagreement with the programmer over whether there should be screen captures in the help or not.

The next day I went to the R & D department to ask a few questions. The programmer asked me the same question: "Are you going to include screen captures in your help?"

I answered carefully this time. "If you look at the help system you had for the last version, you can see that each page is quite long. You have to scroll down a lot to get just a small amount of information. Why should we make the help so ugly when the window or dialog box is right on the screen in front of you?"

The programmer smiled. "I had this discussion with the other technical writer. I tried to explain to him that some users, like myself, set their monitors to a resolution that doesn't allow layered windows. I can't see the help and the application at the same time." Then she showed me how my help looked on her computer. It was nothing like what I had expected, nothing like what I saw on my own computer.

I was glad I had made a test help to show everyone. This was the first time I had been given the go-ahead to do a compiled Microsoft HTMLHelp system, and I wanted everyone, including myself, to see how it worked before we ignored the disadvantages of hooking up exclusively with Microsoft.

Looking at my distorted help taking up the whole screen for a few lines, I suddenly saw the problem from the programmer's point of view. I saw the problem from the point of view of a different set of users. Suddenly it came to me: a solution. It was so simple, I slapped my forehead and said "Duh!" out loud. My solution was thumbnails. Excited, I explained my idea to the programmer. When I finished, she turned to another programmer in the office and said, "I like her! She actually listens to us."

Even with the excitement of discovering a compromise that would make everyone happy and add to the beauty of the help system as a whole, I was wary of stepping on toes. So, before discussing the idea with the other technical writer, I went back to my test help and included a few test thumbnails.

I took the screen captures that the programmer wanted. I put each full-sized capture on a separate HTMLHelp topic. Then, in a graphics program, I resized each capture to 45 × 45 pixels, the size of a big icon, and saved the resized picture (the thumbnail) under a new name. I inserted the thumbnail into the topic that described that dialog box or window. Then I linked each thumbnail to the topic that contained the actual capture. The result was that the user could ignore the thumbnail and read the text alone or could click the thumbnail to see a pop-up picture of the screen described in the topic. The screen captures had been included in the help system without taking up window space.

Part III

Life On and Off the Job

Technical communication, like other professions, pervades the whole life of the professional. As we so often discuss among ourselves, technical communication is an interdisciplinary field. It is no surprise, then, that technical writers have many intellectual and cultural interests. But those interests don't always coexist easily. In the following five narratives, technical communicators struggle to reconcile their professional and personal lives and how difficult it is sometimes to avoid taking the concerns of one world into the other. Moreover, it has become more evident in recent years, perhaps as technical communication becomes increasingly specialized and professional in its practices, that what we do professionally is misunderstood or not understood at all, not only by the people we work with but also by our friends.

In "What a Life," Christine Pellar-Kosbar struggles to reconcile a number of conflicting loves in her life. Many will acknowledge similar struggles while following her experiences on her commute, at work, in yoga class, and at home. Pellar-Kosbar reviews her career choices, going back in her memory over several diverging roads and the choices she has made, in college and in previous professional jobs. We may understand, with rueful amusement, her compulsion to edit restaurant menus and to convert academic assignments into bulleted lists and procedure manuals.

L. M. Hayes is fortunate in having a fiancé who does understand what she does—he's a technical writer himself. But in her narrative "It Isn't What You Write That Makes You a Tech Writer: A Love Story," she isn't sure she's a real technical writer. As the couple go out on a dinner date, they argue about what it really means to be a technical writer, perhaps clarifying for many of us why this profession is important to us—and why it drives us crazy to find misspelled words in public places!

Jerry Kenney has been a writer for a long time, perhaps longer than any other author in this collection. In "Fluff," he looks back over a checkered career, in which he has been willing to do just about anything as long as it involved writing. Kenney may represent a different breed—or at least a different generation—of writers, but his story reveals something of the history of the profession and suggests that the culture clash between SMEs and writers is not a recent problem in our field.

Clashes in the professional workplace are not necessarily rooted in cultural differences. Reva Rasmussen's narrative, "Madame Mao in the Midwest," portrays the demoralizing effects of poor management, fostering dishonesty and pettiness in a medical documentation setting.

The most unusual narrative in the collection, Lenore Weiss's "Stranger in Paradigm" makes use of modules, an approach to writing that she says she uses regularly in her professional work. Here, she uses it to construct what she calls a collage of images, impressions, and both internal and external dialogues. Her story explores the process of creating meanings, whereby people construct—and deconstruct—their world and their lives. Her narrative is a delightful and disturbing postmodern view of contemporary existence, questioning our conventional notions of time and space, continuity, work, and reality itself. After reading this story, you may never again see an elevator, a cell phone, or a cup of espresso as you did before.

Christine Pellar-Kosbar

What a Life

Christine Pellar-Kosbar lives in Ann Arbor, Michigan, with her husband, Kyle, and daughter, Meara. Christine is a technical writer for the University of Michigan, documenting Internet routing software. In her dwindling free time, she practices yoga, occasionally writes fiction, and studies vegan nutrition, conflict resolution/noncoercive parenting, and religions.

"You have to make a decision," said the little voice in my head. "As a hobby, grad school is too time consuming and too expensive." I was walking from the International Center to the bus stop, passing students on either side. Everyone else was paired up, laughing and talking.

"I don't have to decide today." I sat down at the covered bus stop and rummaged through my green canvas briefcase for something to eat. I found a margarine container of roasted chickpeas left from lunch and started munching. Behind me, the old, and yes, ivy-covered Georgetown Administration Building cast a shadow over me, making it just a little too cold to leave my coat unbuttoned. Still pretty warm for January compared to my five winters in Houghton, Michigan, but I'd been living in D.C. for nearly three years and lost my resistance to cold. I buttoned up my coat and hunched into a corner with my chickpeas. The students swarmed all around me, hurrying to class in their long skirts, tailored blouses, and suit jackets. They seemed so much younger, so much hipper than I ever was. Students at Michigan Tech seemed more like me. Maybe it was that so many of us at Tech had working-class parents, or maybe that under a ton of winter clothes, differences don't show up so much.

I looked down at my clothes. Under my black wool dress coat, I was wearing a navy blue pleated suit skirt, a white cotton blouse (now untucked), blue tights, white sweat socks, and black walking shoes. After work this morning, I hadn't had time to change clothes. In fact, I had run from the subway to make class.

Becoming a technical editor is like getting a disease. You can't read a restaurant menu without seeing a typo or noticing ways in which the format could be improved so you could find the soups. You try to go back and read a professional journal article, and you are just not willing to wade through the jargon unless you're going to really get into it and rewrite it. You become afraid to write anything yourself, sometimes afraid to speak, because you hear and see mistakes and know that you can't find them all. Without the ANSI Style Guide *or the* Chicago Manual of Style *at your side, you feel just a little lost.*

"Eventually," my voice said, "you have to choose between school and work."

Work. I popped a chickpea and found myself organizing my projects while watching tiny snowflakes blow around in the air, melting long before they reached the ground. Work right now means preparing for the big Show my company puts on. I'm a technical editor for a nonprofit trade association. We put on a big trade show each year for the imaging and information industry—that's *imaging,* as in scanners, microfilm, and document management; not *image,* as in "dressing for success." This year the Show would be in San Francisco, and we only had a few days before our equipment had to be shipped. My projects for Show were nearly all done because most had to be to the printer a month before ship date. But I was still designing a set of templates for the volunteers who wrote the national standards that I edited. The style guide I'd written a year ago wasn't being used as much as I'd hoped; they wanted templates instead.

Templates meant using Microsoft Word—what a drag. I'd been using FrameMaker for the last year. But if I made the templates in Frame, most of my volunteers couldn't use them. Bleah. I'd have to switch to Word. Can't make volunteers buy Frame. Wish I could.

Maybe I could do some sort of translating back and forth? No, the whole point of the style guides and templates was to minimize my formatting time. Translating would just slow me down. I needed a way to go faster. Twelve committees fed me their standards and expected me to edit, format, resolve inconsistencies, and track down graphics. I didn't have time for translation problems. Oh well, in Word I had some nifty macros that would speed things up a bit. But I would miss Frame. I'd miss the way my text stayed on the page where I put it. I finished my chickpeas and tossed the container back into my briefcase.

I stood up as the bus came around the driveway for the Georgetown hospital and stopped. I flashed my ID, found a seat next to a sleeping student, and sat down. He looked so much younger than me. At most he was, what, seven years younger? If he was a freshman. I'm not that old—why do I feel as if I don't belong here even after a term?

"You don't belong here," the voice in my head said. "Your term was one class. Everyone you went to class with is now immersed in academia, in each other, in different classes and ideas. You have been merely moonlighting. Your real life is still technical editing—getting that question resolved, making sure that standard has been balloted correctly, making sure the printer has the EPS file, checking that the figures match the text. These classes have been just a game here. You have to choose. Either go to school full-time or give it up."

I pulled over my briefcase and looked at the assignment for the next class. "Deictics—Giving Directions." Oh, no, deictics. "Ask three friends to give you directions to their houses. Tape-record the directions, and compare the way they use the terms *this, here, that,* and *there.*

At Tech, I had studied power markers between teachers and students—here I'm supposed to study how people use the word *there.* You know, *there* is a book. *There* is the French Embassy.

I looked at my assignment. You know, if this were part of an ANSI technical report, it would look as follows:

4 Requirements

For this assignment, you will need a tape recorder and three friends.

(Wait, would it be "3" friends? Are "friends" a quantity being measured? If so, you have to use Arabic. I wonder if you'd need to define "tape recorder" up in **3 Definitions,** or if ANSI would let that slide.)

5 Requesting Directions

For each of your friends, begin recording with the tape recorder. Ask the friend to tell you how to get to his or her house by car . . .

(Or perhaps a numbered list of steps would be better. Depends on the audience—would they be offended by a discussion of recording methods?)

Becoming a technical editor is like getting a disease. You can't read a restaurant menu without seeing a typo or noticing ways in which the format could be improved so you could find the soups. You try to go back and read a professional journal article, and you are just not willing to wade through the jargon unless you're going to really get into it and rewrite it. You become afraid to write anything yourself, sometimes afraid to speak, because you hear and see mistakes and know that you can't find them all. Without the *ANSI Style Guide* or the *Chicago Manual of Style* at your side, you feel just a little lost.

"So drop it, go back to school as you planned."

Not so fast. On the other hand, not only is the money great (and I work for a nonprofit, where money is not usually great), but I have nearly complete control over my day. I get to work from home whenever I want, which is handy when I need to get some actual work done. And I have nearly complete control over what software I'll use—well, except when the volunteers need templates. And I got to go to Boston, New York, and Seattle for classes, conferences, and Shows. That's a lot of fun. I've been to every truly great vegetarian restaurant that I've heard of. And I like having a tangible result of my work. Last week, I edited a standard that shows you how to avoid poisoning the water supply when using silver in microfilm processing. That's kind of cool—helping people avoid poisoning their town. I also did a technical report that explained a method of statistical sampling you can use to test the quality of scanners. It's really fun when you find an error in math or computer code; makes you feel that all those classes at Tech didn't go to waste. Heck, for that water effluent standard, I brought up stuff from one of my thematic clusters—a civil engineering class on water pollution I thought I'd never use. And soon, we're supposed to start on the international standards for the committee on microfilm, which means I'll get to edit in French, finally using my language certificate. How cool is that?

The fact is, I like my job.

Not only is the job fun, but it's only forty hours a week, except right before Show, when it's fiftyish. Usually, I go home at 4:00 and that's that. I take yoga on Tuesdays; my husband and I work at a cat shelter on Saturdays. I'm vice president for the MetroBaltimore STC and have organized a bunch of meetings. I meet with a writers' group once a month and have one novel done and another started. My husband and I go to a science fiction book club once a month, too. We have time to do all those things because we work overtime only occasionally.

The last time I was in academia, school was my life. I worked seventy to eighty hours a week all the time. Can I go back into it and not lose myself? If all my time is spent jumping through hoops, won't I miss all my projects and clubs? Or will there just be different projects and clubs?

The bus stopped. I stepped off into the street and walked to Dupont Circle, trying to avoid the cars and other students hurrying to get off the bus. The cars were already gridlocked, and I had to pick my way through.

"UM-brellas, UM-brellas." On the street next to the subway escalator, an old woman sold scarves, and a young man sold umbrellas and T-shirts out of a huge cart.

One of the students stopped to buy an umbrella. Here, snow usually turns to rain.

I stepped onto the frighteningly long escalator down to the subway and held on for dear life, trying not to look down. It got darker and colder around me as I descended, nearly alone today. Not many people riding the trains yet, still too early.

I walked through the white tile station and down the flight of steps to the train platform.

"At one class a term, you have a decade of hoops to jump through before you can study what you want. And what for?" I heard myself ask as I stood on the yellow line waiting for a train.

But wasn't that what I wanted? Wasn't that what I'd planned? I got my last master's in a little over a year—speed of light. I felt rushed, with no time to reflect on what I was learning, what I was doing. I'd gained fifty pounds, some of which I was still trying to lose. I had wanted to take this degree slowly. I'd wanted to go to school part-time or minimal full-time and get my doctorate while raising a family. I'd wanted to do it slowly, to do it right this time.

"But a decade of meaningless hoops? For a very restricted field? What about the other things you'd like to do? Did you hear what your teacher's life is like? She commutes to D.C. from California! Every week!"

My prof had tenure at Georgetown; her husband had tenure at Berkeley. She flew each Monday night out to D.C. and back to California on Thursdays. I wondered if she got a discount on flights; she'd have to, wouldn't she? What a life, to be separated from your kids day and night, four days a week. What a life.

What a life, yes, but is the corporate world any better? When I was leaving Tech, master's in hand, one of the other grad students had said to me, "You'll hate it in the corporate world. They take everything from you; drain you; work you sixty to eighty hours a week, and then lay you off when you burn out. It's all politics and meaningless drivel. You belong in academia."

"But that isn't true now," a voice cried. "We're happy now."

"But what about when this job ends? All jobs end eventually. What about the next job?"

I had to admit, my first technical writing jobs were awful. If all the other jobs suck, I couldn't continue as a technical communicator.

The light shone on the walls of the subway, then the train was in front of me—stopped. No one here today. They're out driving at 4:00, but not on the subway yet. I got a seat to myself and got comfortable. The train began moving, and I watched the posters on the subway walls begin to fly past my window.

My first job out of college was writing hardware and software documentation for a company that made cash registers for restaurants, hotels, and airports. They paid very little. The staff was generally untrained, unhappy, and transient. Perhaps at the interview, when the human resources manager bragged that they had never replaced their carpeting, now dirt-colored and worn through, I should have realized this company was penny-wise and pound-foolish. But I had needed a job, or at least I had thought I needed a job.

I was amazed to learn how much the client business affected the company. Restaurant work is usually slow with sudden rushes. No one at this company seemed bothered that we sat around doing nothing for months and then ran like hell the days before a release. Restaurants have little profit margin. This company was obsessed with saving money, to the extent that they lost their engineers regularly and hired people who had no training in their fields. I was the only person trained in technical writing.

Actually, I began to learn how common it is for people to change careers, especially into technical writing. On this team, one person had been a teacher, another an artist, another a published novelist, another a musician. At lunches we would walk our corporate campus and then out to the dirt roads and farms behind it, and I would hear how their careers had meandered. My own life seemed so dull and straightforward—and short—compared to those of my coworkers.

Their lack of training did not mean they lacked passion for the field, but did lead to some odd choices of priorities, from my perspective. Our staff meetings would frequently end in screaming matches over the use of bold versus italics in document headings, yet when I asked about user testing and audience analysis, their eyes glassed over. They knew their audience, I was told. They didn't need to spend any money on user testing.

However, they didn't mind the wasted money and time in changing the style guide daily. Every morning, someone would go to the boss, who was unbelievably insecure about his position in the company, and tell him the style guide had to be changed. We'd get an e-mail and spend the next few hours reformatting our work. The next day, or next week, the style guide would be changed back when someone from the other camp trooped into his office and shut the door.

All this might have been fun, and funny, except for the undercurrent of tension. One coworker was *very* unhappy (as opposed to the others, who were merely unhappy). People worked for this company because they didn't want to get top-secret clearances—in D.C. most jobs require them—or because they didn't want to commute into the city; this company was outside the beltway. People did not work for this company because they liked the company. Grumbling and petty sabotage (as I always suspected the style-guide changes were) happened constantly. Eventually, one coworker began following another home and then making threats about the boss in my cube. Maybe now that I've been in the work world a couple of years, I would know not to take it seriously, or maybe I was right then. I don't know. I reported him to our boss, found another job, and got the hell out after only seven months.

The train stopped at Union Station and two women got on board, not together, one dressed in full executive regalia, the other in a long skirt and sweater. As always, they chose seats as far from each other, and from me, as possible. That's the city.

My second job had been out in the suburbs of Baltimore. By the time I found it, I had already given up on industry and applied to Georgetown. I applied early in hopes of getting financial aid and found myself accepted a year earlier than I'd planned to go, but with no aid. I decided to wait the year, try to save some money, and try again for aid. So I started the second job, at a company I thought was the exact opposite of the first. Perhaps when my coworker attempted suicide that first week, nominally over the decision to delete a screen shot from the opening page of each chapter, I should have guessed this would not be the greatest job in the world. But I stayed with it, hoping somehow it would work out.

They created accounting software for attorneys. Just as the restaurant industry affected the company making tools for it, the legal industry affected this company. Everything was beautiful. The office was decorated in pale pinks and smoky grays, as opposed to the dirty blue of my past job. The documentation was professionally bound with pink plastic covers and tabs to help you find your section, not the in-house bound 8.5" × 11" whatever-came-out-of-your-laser-printer of my past job. The documentation was done on Macintoshes with PageMaker, which I thought I would prefer over the UNIX FrameMaker I had been using. I was wrong there. It took three software programs on the Mac to get the screen shots, doctor them up, and get them properly placed in the manuals, partly because PageMaker had such limitations at the time, partly because the screen shots had to be fictionalized.

Everything looked beautiful, but the software we were documenting didn't actually work that well. Neither did the company. The technical writers here were not in research and development, as they had been at the cash register company. Instead, we were in the training department. The engineers were not in our department and made it clear they wanted little to do with us. The manager of our department didn't really get along with the engineering manager (or anyone else for that matter), and so communication between the two departments was strained. When we went directly to the engineers to ask for such things as a consistent method of exiting different sections of the program, we were told that engineers were artists and to document it as it stood. Press Escape to exit Billing; type *Exit* to exit Accounts Received; press Control-*X* to exit Invoices. *We* couldn't even keep track of the way it worked.

Add to this mix a truly unhappy manager. I learned a lot there. I learned not to accept a job where your supervisors have never gone to college and are hostile because you have. Actually, I now see hostility of any sort at the interview as a dead giveaway that I should run fast and far. I also learned that miserable people are miserable no matter what you do. And I learned that my writing really was sloppy. In retrospect, I'm surprised so many people had let me slide before. My sentences ran on for days. I rarely checked the style guide for punctuation. I frequently skipped words in my rush to finish. Basically, the work I turned in was in rough draft form. At my second job, I learned to read, then reread, and then ask a friend to read.

I also learned more about what other people have done with their lives and, again, how goals and careers meander and change. My supervisor had been a financial aid counselor at a college (a good resource for my Georgetown application—but I never asked her help for fear I'd be fired if they knew I was going to go back to school). My coworkers had run child-care businesses, had been software testers, psychology majors, art majors—had worlds of knowledge and experience that I lacked. I had gone to high school, then to college in electrical engineering. Then I switched to technical writing courses (writing about electrical engineering and computers), went to grad school, and became a technical writer. Not much meandering.

I was laid off, thank heavens, after a year there. I don't know how long I would have stayed, getting more miserable every day. I had convinced myself that all jobs were lousy, and at least I liked some of the people here, and the commute and pay were reasonable. I did love the lunchtime walk; I'd found a beautiful prep school where I could walk the grounds alone, playing out scenes from my novel. A low point of my life came when I stopped walking and started going out to eat with two of my coworkers who spent lunches slamming everyone and everything. Eventually, I found my way to their shit list and was excluded from the cat sessions. At that point, I decided that industry sucked. The people were about as deep as cookie sheets. When the company began to fail and laid me off, I was completely ready to go to grad school, (sort of meandering into Linguistics) but still had some months before my term started.

I found my present job—my dream job—while waiting for classes to start.

I took the job because the tasks looked different from documenting software and hardware. I'd be editing national and international standards. I thought the variety would be nice and that becoming an editor would force me to improve my writing style. It was close to Georgetown as the crow flies (but still an hour away by subway and bus). And I liked the woman I'd be working for (which, it turns out, is the most important aspect of a job).

"Silver Spring Metro Stop. This train does not continue. All passengers, please exit the train."

I stood up and walked to the aisle, holding carefully to the seatbacks and my briefcase as the train wobbled to a stop. I walked out onto the platform and stretched. All this sitting can get to you. I walked down the stairs, slid my ticket though the taker, and stepped out into the sunshine, now fading behind the tall office buildings. No rain this time.

I worked in Silver Spring; the subway dropped me off practically at my company's doorstep. I walked over while deciding what to do next: Check my e-mail? Go home? It wasn't quite 5:00. (My class had been a 2:30–4:00.) My yoga class didn't start until 7:30. It was almost an hour bus ride home and a half-hour walk from the bus stop. That left an hour. My husband was pulling a series of late nights at work, so he probably wouldn't call home for a ride before 10:00.

"Well, no point going home to an empty house; I'll just check e-mail."

We're on the eleventh floor. I've thought about walking, to burn a few calories, but all the doors are locked above the first floor. You can get out in a fire, but no one—meaning purse-snatchers—can get in. So I took the long elevator ride.

When the elevator doors opened, Rhoda nearly threw me over. Rhoda was a very tall, muscular, dark-haired woman with flashing dark eyes. When she was angry, everyone knew it.

"Hi?" I asked while sidestepping out of the elevator to avoid an elbow in the stomach.

"I can't stand her anymore. Tell that bitch I left." The elevator doors shut.

I turned to see Lynda closing the security gate at the reception desk. Lynda also had dark hair, really curly, but she's pretty short and thin, and pulling down the gate seems to take all her strength.

"Is she okay?" I pointed to the elevator as I limboed under the gate before Lynda shut it entirely.

Lynda shrugged. "It's Show—does this to everybody. And Sherrine isn't helping much."

"Oh. They're at it again, eh?" I laughed.

I walked back to the cubicles to see Sherrine carrying signs to her cube.

"You need any help?"

Sherrine nodded and pointed to a stack of signs. "Girl, what are you doing here after four?" She walked toward her cube.

"My schedule's changed. New term at school. Tuesdays and Thursdays I'll be in early and late around class." I followed with the rest of the signs.

"That stupid designer still hasn't shown up." Sherrine set the signs down in her cube. Her jaw was clenched, as it had been for weeks. She was trying so hard to prove herself. She'd just hired in as the department assistant but was clearly headed for much more—if she'd relax. If she kept pissing everyone off, she'd be chased out. No one ever gets fired here. They had chased out a couple of people in the year I'd worked there, making life so miserable they chose to quit. But no one's ever been fired.

"That fucking designer said she'd be here with sketches a week ago." Sherrine threw herself into her chair.

"Oh, wow. It's only a few days before ship date." I set the signs down next to hers.

"I've called her four times." Sherrine's face was red; she was biting her lip. "She isn't answering her phone."

Sherrine had a reputation for being a bitch, but she wasn't really. I saw her frustration and her new clothes that, like my own, were the least expensive version of the required "work casual" dress code. I could tell from her speech that she was where I had been four years before: a working-class kid fresh out of college, and no one she knew had ever worked in an office. She wanted desperately to do a great job, but she hadn't learned how to fit in. I could relate. Sometimes it seemed that everyone else's parents were university professors or doctors. They seem to know the rules before they walked in. But the rules aren't that hard to learn.

Sherrine really had done a great job and was about to be promoted from departmental assistant to head of an important coalition project. And she had designed a database that made my job about a million times easier, so she could scream her lungs out at me and everyone else, and she'd still be golden in my eyes.

"God, everyone is so incompetent," she said.

I laughed. I could remember thinking such things. "Actually, compared to my last two jobs, everyone here is amazingly competent," I said. "No stalking, no suicides, and practically no management—and everything does get done, usually well."

Sherrine chuckled, her face finally losing that tightness. We drove to work together occasionally, and I had told her about my former jobs. She had listened with eyes and mouth wide. She said she had thought that working with educated people would be different from the waitress and gas station jobs we held through school. But though benefits and working conditions are much better—less physical work because you think for a living—the people aren't much different.

Now she was shaking her head. "But what am I going to do about this stupid, fucking designer?"

"Call her again. Find out what the deal is. Fire her ass if you have to. I know two freelance designers who would help. We could design the booth ourselves. But don't make Rhoda quit, eh? We're going through temps like water here. She's been here a while and knows the routine."

Sherrine shot me an angry look. "I told her not to touch the damn signs."

"Yeah, but I saw the signs leaning on the wall, and that's where you put things that are being shipped for Show. Didn't they tell you that?"

"But I told her not to touch them."

I shrugged. "You've got to get along with people. Relax, learn the rules."

"Fuck the rules; I've got work that needs to get done."

"To get the work done, you need Rhoda. You need Lynda. You need everyone. But, on the bright side, they need you too."

Sherrine laughed. "Rhoda's survived worse than me. I won't chase her out."

"Maybe," I said. "You know, everyone here has designed a booth at some point. We could figure out how to do it ourselves."

Sherrine pointed to a stack of papers where I could see the beginnings of some designs. "Yeah, we probably could."

"All you need is an idea of which supplies to send. We can fill it in over the next month and even carry on some things on the flight out."

Sherrine nodded. "I'll give the designer one last try tomorrow." She shook her head. "Julie said she would be great."

"Julie doesn't usually screw up. Maybe the designer sent the plans by courier, and they got lost. Have you checked the front desk?"

"Yeah, they aren't there." Sherrine rubbed her eyes "Have you eaten? What's good around here? I've only been to that horrible pub right downstairs."

"No, I've got yoga tonight. Not supposed to eat first. We have everything here though—Chinese, Caribbean, North African, Thai, Ethiopian—and that's within walking distance. If you're willing to get on the train . . . "

"No, I think I'll order in. You sure you don't want any?"

I looked at the clock on Sherrine's PC. "No, I think I'll check my e-mail and head home. Kyle's pulling another late-nighter, so we'll probably eat a late dinner."

"Again? Hasn't he worked late all week?"

"His company pays overtime, and we need the money if I'm gonna go to school full-time anytime this decade. Anyway, they're really nice about it. Last time he had to put in a lot of overtime, they took us on a one-night cruise afterward." I started to head out of her cube.

"Sweet. We should do that. Why don't we ever do anything like that?"

I turned to her, surprised. "But we get Show—at Show you get a very decent spending allowance, and we all go to the best restaurants and shows. In New York last year, I saw three Broadway shows. You smile and help everyone out all day and then party all night. And then we have the big dinner/dance the last night of Show. Didn't anyone tell you? You need a formal dress. It's a blast. Our whole department goes together."

"Well, I wasn't going to go."

"Why not? These are the perks. We have a great time."

"I can't afford a dress."

"Everyone here has gone a zillion times—they all have dresses that they won't wear again. Borrow or buy one. Who's your size? Hmmm . . . I'll bet Rhoda is. She's been a temp here for six years; that's six dresses to choose from."

"Like she's even going to talk to me."

"Good point. How about Marlene?"

Sherrine shook her head. "I don't think she's talking to me either."

"Starting to see a pattern here." I scowled. "I have a little black dress that I wore last year. But I've got to be a couple of sizes bigger than you."

"I'll find something."

"Good! They don't have anything like Broadway in San Francisco, do they? I wonder what we'll do over that weekend after we set up."

"I'm gonna eat."

"Eat! Eat! The Caribbean place around the block is better than the one downstairs."

I walked past the Resource Center where my best friend, Marlene, works. She was gone; she keeps strict hours—seven to four. I went to my cube.

Two more perks about my job: First, because my work had to be done a month earlier than everyone else's, I got to play helper that last month, like helping design that booth. I got a year's worth of favors done. Second, I had a window.

My cube was teal; everything was teal. I had cultivated a long purple and green vine up one side of the cube and over the top along the walls. I held it to the top with push pins. On a large column in the middle of my cube, I'd taped a long poster with quotes about the joys of editing and writing. On one of my two desks, I had a mound of papers, projects in order of when they were due. On the other desk, I had a twenty-one-inch monitor and a PC. I had pictures of my husband and me on our last vacation push-pinned to the cube walls (we'd gone up the coast from Chincoteague to Rehoboth Beach), along with articles about tech editing I really liked and projects that I might forget about if they weren't in front of my face each day. It was a big, beautiful cube, but no one else had wanted it because of the column.

I looked out my window at Silver Spring, a strange little city on the outskirts of D.C. Every day at lunch, Marlene and I walked the streets of Silver Spring, so I'd seen nearly all of it. Near the subway, four huge office buildings stood, but only four. Then there were three blocks of little hole-in-the-wall restaurants, used bookstores, antique shops, and quirky little things you don't see just anywhere, like the jazz and R&B record shop. Then you had houses and schools, all brick, built in the early fifties, with fabulous azaleas everywhere in the spring. The main street had this strange pink, art deco look to the buildings, and a lot of the buildings were vacant—so many that several were being torn down and replaced with grass and flowers. From my window you could see a huge recently planted grass plot, a real diner, a police station, and the huge offices across Wayne Avenue.

Marlene and I had walked up and down the hills of Silver Spring, through the expensive neighborhoods, where she would tell me stories of when she had been a wealthy doctor's wife. We'd walk through the office complexes, and I'd hear about what it was like to have to wear white gloves and matching shoes to job interviews. We'd walk through the neighborhoods, and I'd hear what it had been like to be a law student, to get your library science master's, to raise two kids, to be the head of a library and then get laid off and work part-time as an interior decorator. Marlene had a meandering, exciting set of careers behind her.

Would I, twenty-five years from now, have such an interesting, intricate past? Would I be where I wanted to be?

Where did I want to be?

I booted up my PC and read my e-mail. Nothing earth shattering, although one of my volunteers did send me some figures I needed to get a standard out the door. That would help move things along. I decided to head out.

I walked the long way out, around the Resource Center again, because the front gate was down. A few people were still around, fighting with their computers or packing equipment for Show. I rode down to the first floor and walked out to the bus stop down the street, under the elevated train station, past the huge mural of penguins riding on the subway.

It was now rush hour, so the line for the bus was pretty long, but the buses came every fifteen minutes. The sun had nearly gone now. I buttoned up my coat, sat down on a cement flower box to wait, and found my internal argument starting up again.

It wasn't just that I didn't fit in at Georgetown; I knew that was just a matter of time. It was also the money thing. We were completely out of debt, my husband and I. We didn't have any savings yet, but we were finally out of debt. Georgetown wasn't going to give me financial aid, and my attempts to find it had failed. I didn't want to be in debt again.

In line to register for classes, on a very long lunch from work, the guy in front of me had mentioned that he was over one hundred grand in debt for grad school. He said it as if that were nothing. I had shuddered and felt sick. Even with my company paying for these first few classes, how many years would I be paying for this degree? It was getting so I was constantly nervous during classes because every minute I drifted off, I was throwing away $10.

"You don't have this kind of money, and if you did, would this be the way to spend it?" I had heard myself thinking. "Is this a college, or child care for the wealthy?" Cranky, cranky, but it's hard not to be cranky when you face a lifetime of debt.

The bus pulled through the traffic light and stopped in front of the line. I followed the line up to the driver and handed her my ticket, then headed for the very back, where you can find a seat to yourself and stretch out and go to sleep. Usually, I either read or slept on the bus ride home.

We pulled out of Silver Spring, past the art deco, the hotel that always looks cramped, shoved into a too-tiny spot, out to the churches and houses and then the beltway. After that, the strip malls, and finally, the tree-lined open road of U.S. 29 breezed past. In college, people had told me this was one long city—Baltimore to D.C.—with no space to breathe. That wasn't true. There were heavily developed areas—Laurel, city of malls, for example—but on 29 and the parkway it was lovely and quiet. At both my former jobs, I could easily walk either out into the country or out to a major highway. There was city, but it wasn't all city. It was like that overgeneralization about working—there indeed were companies that would work you to death and then lay you off, but they weren't all that way. Each company has its own personality; you can't make generalizations.

Same with colleges. Georgetown was not Michigan Tech.

Those inner voices started in. "We can't keep going part-time. We're not really part of the department. We can't live a life on buses and subways. If you're going to do it, go full-time."

"But I don't want the debt; and I'm not even that hot on Linguistics as a career. This stuff on deictics—this isn't exactly Tannen and Pinker here. This is starting at the very beginning. This is years of hoops."

"Well, what did you expect? To just walk into Georgetown and start with your own research assistantship?"

I had, actually. When I started my first master's, I already knew my school. I knew the ins and outs. I got a teaching fellowship and taught composition and technical writing classes. I thought Georgetown would be like Tech, but better because I had done real work. I really thought that somehow I'd be worth more—I could teach a technical writing course now with real-world experience. I thought they'd see all I'd done on my own and find a place for me in sociolinguistics or computational linguistics.

"It will take years. No one knows you here. You'd have to commit first, find out later if you could work something out. Meanwhile you're just another student. And you have to really like this stuff, as you did at Tech, to do something worth noticing."

Did I really like linguistics? Enough to override my interest in tech writing and everything else?

When I was still an undergraduate at Tech, still an EE major, I'd look over the catalog and feel sick. The classes I was set to take held only the slightest bit of interest to start with and none at all after I'd taken a couple and learned the nasty secret of engineering schools: with a couple of exceptions, the profs can't teach. They are there to do research, and you'd best get yourself a kickass study group and teach each other. I'd turn the pages over to the humanities and tech writing classes, and I could feel the nausea subside. I couldn't help liking classes in which you did more than memorize equations and try to beat the curve; classes in which the prof spoke *to* you, instead of *at* you, and expected you to speak back and know what you were talking about. I'd tried to figure out how to fit those classes in until, finally, it became obvious that the best way was to change majors.

Now, when I looked through the classes I would be taking, I felt some interest, some pull, but was it enough?

One of the women in my writers' group, a technophobe who wrote fabulously beautiful physical descriptions, was finishing up a Ph.D. When I told the group of my plans to go back to school, all she could say was "Why the hell for?" Stories of horrible committees, of rewrites and funding shortages followed. My experience at Tech had been so good, I had not realized it could be anything but good.

"Why get the degree?" another writer had asked, "If you like linguistics, study it on your own. You can study what you think is important."

I had wanted to teach, but her point was valid. I'd studied on my own for the last three years, and not just linguistics. I'd studied conflict resolution, nutrition, child-rearing practices, religions—all without anyone else's hoops. Why go back? Why not move ahead? I could do a lifetime independent study on topics I care about.

But it could be so exciting, doing research again, discussing ideas, teaching. And this time, I might eventually get past the drudge work of research and on to the interesting part. Less time poring over manuscripts and searching for politeness markers and more time analyzing.

"A decade from now you'll be into the interesting part, maybe five years."

The bus pulled into a Park and Ride, and I stood to prepare for the next stop. We were in the city of Columbia, the village of Owen Brown, my home. The bus pulled into the village center, and I stepped off with four or five other commuters.

Columbia is about the most beautiful place I've ever been to. I used to joke that I took all my vacations back to Detroit because I had no need to visit somewhere gorgeous where I knew no one—I lived there. Columbia is a planned community, one of the first. A network of walking trails could take you anywhere: one of the several parks throughout the town or one of the two brand-new, gorgeous, constantly busy libraries. Everything was clean and quiet. Each neighborhood was built around a village center, a set of small stores designed to put everything within walking distance: the grocery, the bank, the beauty shop, the interfaith center. Each village center was decorated with big wooden flowers or orange and yellow swirls, unmistakable sixties design. I felt guilty sometimes at yoga when the others talked about how wonderful Columbia was before the yuppies came. I knew they saw me as one of those newcomers who spoiled the plan.

I walked past the Hardees on one side and Lake Elkhorn Park on the other, past the Giant Grocery and the dark wood condos. The beat of walking lulled me into a calm.

I had realized, months earlier, how much I really liked this job because I would daydream while walking home, not about the next chapter of my novel, but about the next section of the style guide. At my previous jobs, I did everything I could to avoid thinking about work, even while I was at work. But I found myself thinking about this job, happily, anywhere I went.

So now, as I walked, I tried to think about school. Last term I had taken one class, a required introductory course that had really been an in-depth version of a class I took at Tech. I even used the same book. The other students had been friendly at first, but by the end of the term they knew each other from other classes, and I was the odd one out. I did try. Kyle and I had gone to the departmental holiday party and tried to strike up conversations. But not being there full-time made it hard. The class was easy, and only one night a week for three hours. I did the assignments on the bus and train, and rarely had to find other time. It was a fun hobby, but nothing serious.

This term was different. The class took two nights out of each week, and it looked as though the homework and readings would be extensive. I got the impression my boss would let me get away with not making up the actual hours I was in school, but there would still be some juggling to get all my work done.

"Is it worth it: the money, the time? What are you going to do?"

I sighed. "Go to yoga."

My neighborhood, Sewell's Orchard, was really an orchard until ten years ago, when it was bought and turned into a subdivision. The old-timers won't let you forget that—that you ruined their local orchard.

My townhouse was tiny, but cute. When we drove up to it the first time, I didn't even want to go in. I thought it looked too yuppie, too Stepford with its blue aluminum siding and tan trim. You could almost hear the real estate agent mantra: "neutral, neutral, neutral." I wanted one of those old brick townhouses in the "bad" part of Columbia, but my husband didn't like their dark, oppressive rooms. He'd wanted a single-family house, but I didn't want the extra yard work, and we couldn't find one in our price range that was designed well. We reluctantly went in to this one and both fell in love. Big mistake, you know, falling in love with a house. We probably paid too much for it.

I unlocked the door and ran up the stairs, pulling the blouse off as I went. I grabbed my sweats and got out of the skirt and tights. Fifteen minutes to yoga.

Kyle had ridden his bike to work so I'd have the car. I flew back downstairs and out the door, jumped in our little eco-box, and started out into the confusing, twisting maze of beautiful Columbia roads. It's easy to walk somewhere in Columbia, but it's damn near impossible to drive. They designed the roads with hills on either side, to hide the development. It's lovely, but you can't figure out where you are. Everything looks the same. Kyle jokes that when people pull over and ask him for directions, he's always tempted to say, "Slide over and let me drive you there. It's easier to drive over and walk back than explain it."

At the unusually long Columbia stop light, I heard my voices again.

"We have worked for years to finally get into Georgetown; now after only a term, you're so ready to give up?"

"Well, it isn't what I expected."

"But it might be. It might take a little time to find your place, but you always do. Why not give this a real chance? Quit and go full-time. Fall back into the excitement of research and all the new people, new ideas."

"The debt, the time commitment, the lifetime of self-centered, self-absorbed careerism, I just don't know. You have to be obsessed to enjoy academia, at least at first."

"But what choice do you have? Where are you going when this job ends?"

My first day at the association, one of my coworkers, a hysterically ambitious office manager who started the week before me, said, "You know this job is dead-end. You aren't going anywhere from here. Why did you take it?"

I looked up, not quite believing she'd actually said that, and replied, "Because I want to be a better writer and editor. I don't want to be a manager, except a manager of a tech writing department maybe. I know there is no tech writing department here. I know I'll have to move on someday."

She was visibly relieved and actually turned out to be a good friend, once she realized I was not a threat. I had thought I would move on to grad school, of course. Having pretty much given up on industry, I'd pretty much given up on managing a tech writing department.

I pulled into my yoga teacher's parking lot.

"So what do you want to be? Linguist or tech editor?"

When I was thinking about applying to grad school the first time, my advisor had sat me down and said, "Where do you want to be in five years?"

I believe I stared at her without even a hint of expression because she laughed and said, "Sometimes this helps: what do you think you'll be wearing?"

I looked down at my blue jeans and thought, What do I want to wear? What kind of question is that? I made up what I thought an academic would wear: a long brown cotton skirt and a vest. Now I wondered if maybe she was on to something. What did I want to wear five years from now? Who did I want to be? Did I want to be wearing the cheap skirts and blouses I wore now? Suit? Jeans? Yoga leotard?

"Who do you want to be? Where do you want to be?"

My yoga teacher, Mark, was standing at his doorway, letting in the students in sweatpants. He was an older man with a full gray and white beard. He's one of the nicest people I've ever met. One of the things I love about Kripalu yoga is that the teachers are so uncritical about people.

"Are we done here?" I asked myself. "No arguing in yoga."

I got out of my car, my mat rolled under my arm, and smiled to Mark. I walked inside to his empty living room. I've never seen furniture in it and don't know if he owned living room furniture. He had straw-colored, braided carpeting, and the house smelled of muffins. Mark always gave us a muffin and lemon water after yoga.

We're not supposed to eat beforehand, so even though he didn't usually use sugar, they tasted terrific.

I unrolled my mat, trying not to crowd the woman next to me. Celia, I think, was her name. There was room for eight easily, but on crowded nights, we'd fit twelve in that room. Kripalu yoga focuses on relaxation rather than strength building, flexibility, or balance, although those are important too. You hold a pose for a long time, relaxing into it, focusing on the breath, always the breath. You wait for the stretch to become too comfortable, and then you stretch a little more, breathing and controlling, relaxing all the muscles not involved in the stretch, stretching only those that are.

You can't think about anything else and still do yoga. No work, no grad school: just inhale, stretch, relax, exhale, relax.

Class began, and I found myself falling into the routine. I'd been taking yoga with Mark for over a year, and it was about time to change teachers. I knew these poses too well. I didn't have to concentrate so much anymore. I could feel the grad school decision still pressing on me. Pushing me down.

The last posture in any Kripalu class is Shivasana, the corpse. You lie face up, shoulders back, eyes closed, under a blanket if it's at all cold. You control your breathing and begin to float. Once you learn how to get to that floating feeling, it's easy. You can stay and think, or not think.

And suddenly, I realized I had more than two options. There are more than two options. I don't have to choose between grad school now or a crappy corporate tech writing job later. I can choose to go back to school later, in linguistics or anything else. I can choose to freelance as a technical writer later. I can choose another nonprofit tech writing or editing job later. I can choose to teach yoga later. I can choose to teach tech writing later.

I love my job and my life right now, and there's no reason for me to give it up.

Inhale, exhale. Relax.

L. M. Hayes

It Isn't What You Write That Makes You a Tech Writer

A Love Story

L. M. Hayes, a writer since the sixth grade, lives by two hard-and-fast rules: All writing is technical writing. All writing is creative writing.

It happened again. I've known Julie for some time, but only casually. We were coming out of church after a meeting one Saturday afternoon, and after I mentioned something about work, she asked, "So, what do you do for a living?"

"I'm a writer," I answered.

Her response was the same as everyone else's always is. "Oh. What do you write? Books?"

As a matter of fact, I'm a technical writer, but if I had said that, Julie would have asked one of two questions: "What's that?" or "Oh, so you write software manuals, or something?"

Not that I mind explaining what technical writing is, but I prefer to save time when I can. So instead I gave Julie my standard response. "No, right now I work for a marketing firm as a copywriter." Most people have a basic enough understanding of the term *copywriter* to let it go at that. Even if their understanding is based on Darren's job on the TV series *Bewitched,* they tend to accept that I am the person who comes up with all those catchy slogans and jingles.

> *"I am constantly amazed at how quick most people are to blame themselves when they don't understand something explained in writing," I went on, hoping I wasn't raving yet. "I always hear people say, 'I don't understand this. I guess I'm just not a technical person.' Bull! If the directions are there, they should be clear enough for the user, any user."*

As a matter of fact, I don't write slogans or jingles, but it's close enough. I don't write software manuals, either, or anything else that most people would consider "technical." The primary business of the marketing firm where I work, however, is very technical, having to do with software and huge information warehouses that clients (mostly in health care and financial services) use to market their products and services directly to customers. Direct marketing doesn't sound very technical, either, and I'm in the marketing support area, so I don't even have anything to do with the real technical side of the business. We have a documentation specialist for that, and I only proofread manuals before they go out.

After saying good-bye to Julie, I went home to my best friend and fiancé, who happens to be a "real" technical writer. He was lucky enough to get a gig with a big insurance company as a bona fide technical writer. It's on his card and everything. I'm so jealous.

When I got home, Tim was on the computer.

"Working late?"

"No," he said, "just noodling with this new software package. It does twice as much as the old one, but it isn't very user-friendly. My team really has our work cut out for us."

I sighed, thinking, This is what we learned in grad school. This is what I should be doing.

"Something wrong?" asked Tim, always the perceptive male.

"Nothing, really," I answered, which in relationshipese means, you bet there's something wrong and it's up to you to find out what it is. If you love me, that is.

"What is it?" he said. I must say this about Tim. Considering that we've been together only four years, he was really starting to catch on.

"Oh, it's just my job," I said. "No big deal, really."

A lesser man would have taken the out, but not Tim. He actually turned off the computer and turned toward me. "What is it? Your boss again?"

"No, that's going fine, really. I just don't feel that I'm doing what I went to school for. I'm not a real technical writer like you. I write letters and brochures. Anyone can do that."

"Anyone? What about all the client-supplied crap, I mean the copy you get and have to make readable?" he asked.

"That's just because they let doctors and nurses try to explain what they do, and they know it too well to explain it to Joe Consumer."

"I see," said Tim, pensively. "And when I finally figure out this new software package, after the programmers have to explain it to me, and then I have to make it understandable to Jane User, how is that different?"

So you see why I'm going to marry him, don't you?

"Do you still want to go out and get something to eat?" he asked. "We can talk about this on the way."

"Sure. Just let me change into jeans."

Ten minutes later we were on our way to our favorite rib place. Tim, being the "real" technical writer, never loses his train of thought once he's on a job.

"So you don't feel like a real technical writer because of what you write," Tim began. "Maybe you *didn't* learn anything in school," he teased. And he smiled to let me know he didn't mean it. "Instead of trying to convince you myself, why don't you make your own case that you *are* a real technical writer?"

"Hhmph," I said, or something like that. I had been, of course, looking for blind moral support and a little ego stroking, not a basic review of tech writing in the form of a modified high school debate. But Tim and I always love to argue, even when we actually agree on a topic. It's one of the things that brought us together in the first place. We've always had a way of bringing out each other's intellect.

"You want me to prove that I am a tech writer?" I asked. "Okay, what makes me a tech writer? Well, for one thing, that's what my education is in."

"Good start, but we know lots of people who got degrees in things they'll never do for a living."

"Okay," I continued, "what's more important is how I approach writing. Every time I write copy, I put myself in the reader's head and figure out how I can make that person understand what I'm trying to say. I try to make information accessible to the person. The information may or may not be technically complicated, but my job is to adapt it to a specific user or audience."

"Well," Tim teased, "at least you haven't forgotten the basic definition of technical writing, 'accommodating a technology to a user.' Have you convinced yourself that you're a technical writer yet?"

"Maybe. But it's still different. If nothing else, because it's a direct marketing piece, I need to move that reader to take some action. That isn't part of the definition of technical writing."

"No," Tim agreed. "But I think that's an even more complicated and difficult goal," he said, pulling the car into the restaurant's parking lot. "I mean, take the user manual I'm working on now. That's what you'd call a traditional technical writing piece, right?"

I nodded.

"Well, in that situation, the reader has some motivation to understand the writing. If he or she wants to learn how to make the tool function the way it's supposed to, then he or she had better read it, understand it, learn it. If, on the other hand, that same person gets a piece of direct mail that you wrote, what's the motivation to put any effort into it?"

"True," I agreed, unbuckling my seat belt. I got out and looked at Tim over the roof of the car. "I'm glad you aren't going to make my argument for me." I smiled sweetly.

"Touché," said Tim, bowing to acknowledge my point. "Continue, please." You can see how much we both love getting into the heart of an argument.

"Okay. What if I concede that I am, indeed, a technical writer, despite my job title and business card? My point is still that I am not doing all I learned in school." We crossed the parking lot toward the restaurant entrance.

Tim held the door for me, and we gave the hostess our name. "There's about a fifteen-minute wait," she informed us.

"Great!" we both said in unison. She looked at us a little strangely, but we just took a seat in the waiting area. We had an argument to finish, after all.

"For example," I continued, "look at all the page design classes I took in grad school. Those were my favorite classes. None of them even applies to my current job, because we have 'real' designers, people from the art department with degrees in graphic design."

At this point, I realized I was on the verge of whining, but I didn't care. "So not only am I not a real tech writer, I'm not a real designer, either."

"Now you're just whining," Tim said. "You don't care about being a designer or an artist. You're a writer. Now, make your argument and quit going off on tangents."

Hmmm, I thought. Married life was going to be more interesting than I thought. I didn't know mind reading was part of it.

"Fine," I sniffed. "Well, in my line of work, the first obstacle is getting the person to even open the direct mail piece. Assuming they do open it, I then face the challenge of (usually) conveying some complicated information. Sometimes it's explaining a new medical service now available in the area and why they should call and schedule an appointment for it. Sometimes it's more general, like introducing a person new to the area to a hospital's programs and services. Again, most people would not consider this information technical, but it is difficult for the reader to understand, even though it's more general. I mean, most people don't understand the difference between a Level I trauma unit and a full-service emergency department. My job is to make them understand that whatever the hospital offers, or whatever the service is, it's good for them and they should do this, or take that action."

Tim crossed his arms over his chest and nodded, which was his "deep in thought" pose. "And this doesn't seem like tech writing to you? This isn't what you went to school for?"

"Okay, so some of it is," I admitted. "But what about the page design and visual element classes we took? And don't call this a tangent, either. Those are really important to the writing, too."

"And the classes you took don't help you at all?" He sounded skeptical.

"Well, I don't do the design, but I have to evaluate it during the creative process. We do work as a team, the designers and the writers."

"I see," said Tim. "So tell me two things you learned in your design and visual classes that you have used recently, at work, in the process of working with the designers to create a document."

"You sound as if you're giving a lecture in a basic tech writing class," I groused. He cocked his head and raised an eyebrow, meaning answer or give up the argument, so I answered. "Dean had the basic *Z* backwards on a piece last week. It started in the upper right-hand corner instead of upper left," I began.

"Dean? I thought you said all the designers there are women."

"They are, I was talking about Deanna. Sheesh. Jealous much?" Inside, however, I smiled.

"Oh, yeah," he stammered. "I was just kidding. And example number two?" He seemed in a hurry to get the conversation back to the argument.

"Same piece," I answered, thinking equally about the document and what I was going to order if we ever got a table. "The content of the inside panel continued to the back, so we invaded the margin in the lower right corner to draw the reader's eye to the other side. The client won't go for an 'over, please,' so we have to make it natural for the reader to turn the page over and keep reading. That was a good example of too much information in too small a space. But what can you do?"

"What, indeed?" sighed Tim. "I hate to be the one to tell you, my love, but you sound like a technical writer to me." At that moment, I wanted to kiss him, so I did. Also at that moment, the hostess walked up to tell us our table was ready.

"Or do you need a minute?" she smirked.

After ordering, Tim picked up where he had left off. "You forget, I live with you. I see how you process information and react to things every day. Being a technical writer is no more about what you do for a living than what you write."

"How so?" I asked.

"Well, have you ever read something without evaluating the effectiveness of the message, to say nothing of the correctness of the presentation?"

"Of course not," I answered. "You know how much it annoys me to find typos everywhere. The newspaper, books, outdoor advertising, even on television. It drives me crazy to find poor writing anywhere. That's the most basic level of communicating effectively."

"Oh, really?" he asked innocently. "Explain."

"Rules about spelling and punctuation, grammar and usage, those aren't for the writer; they're for the reader. And writers of any kind—technical, marketing, fiction, journalism—are required to write in the most correct, clear, and effective way possible."

"I see," Tim said.

"Besides being correct, writing must, by definition, be clear. How can you hope to communicate, which is the purpose of writing, if the writing is unclear or ineffective?" Since that was a rhetorical question, I didn't expect Tim to answer, but he nodded his head appropriately.

"I am constantly amazed at how quick most people are to blame themselves when they don't understand something explained in writing," I went on, hoping I wasn't raving yet. "I always hear people say, 'I don't understand this. I guess I'm just not a technical person.' Bull! If the directions are there, they should be clear enough for the user, any user.

"My response to those people is usually, 'Maybe it just isn't explained clearly enough in the directions (or manual, or whatever).' It can be a totally new concept to someone that maybe, just maybe, the person who wrote the explanation has some obligation to make the reader understand. Back me up here, Tim. If you hand out a user guide to a bunch of people, and even one of them says, 'I don't understand this part,' do you tell them they aren't technical enough, or do you go back to what's written and evaluate it all over again?"

This time I wanted an answer, but Tim just nodded again.

"Well?!?"

"Oh, sorry," he answered. "I go back to the part that isn't clear to them and try to look at it from a fresh point of view. Did I assume some knowledge that wasn't there? Did I not give a clear enough example? Did I skip a step because I know it goes in a certain spot? And so on."

"Yes, yes. Exactly. In my job, I know it's my responsibility to make the reader understand. It's all we ever talk about in evaluating the effectiveness of a piece. Will the reader respond in the way we want? Will the reader immediately understand the call to action? Will the person who receives this in the mail be motivated to open it? It may not sound like the traditional definition of technical writing, but at the most basic level, nothing is more so."

"Really?" said Tim with a smile. "Sounds as if you've made your point. I know I'm convinced."

"You should be. And look, here comes our food."

Jerry Kenney

Fluff

Jerry Kenney has been around so long that most folks expect to see his name followed by bracketed dates. Tracing his steps back through Chicago, Indianapolis, and Buffalo, the common response to the mention of his name is "Oh—Is he still alive?" The answer is yes, still, and working in Orlando as a senior quality assurance specialist for the Gartner Group Interactive Channel Software & Service development team. You can say he is a software tester.

He is also and finally picking up the thread of education he had dropped after leaving the State University at Buffalo with a B.A. in 1964. He expects to earn a master's degree by 2003 and his Ph.D. before his seventieth birthday.

To place him in a historical context, a few months before the U.S. entry into World War II, Jerry was born in Lockport, New York, home of the fire hydrant and the gumball. It is also home to presidential candidate Belva Anne Lockwood (defeated by Cleveland) and vice presidential candidate William E. Miller (defeated by Humphrey).

You might call him the accidental writer. Sheltered from harsh reality by Buffalo's richly exciting literature program, he never knew that the world of commercial writing existed until he answered a help-wanted ad for a copywriter in academic public relations. Two things appealed to him about the job: it paid slightly more than the offer he had from the auto parts store, and it was clean work. Since then he has worked in advertising, sales training, and that vast amorphous domain they call technology. Fields in which he eventually became expert include retail sales, banking, insurance, fertilizer, and telecommunications. Those employers and clients he misses most are the few nineteenth-century companies that did not make it to the twenty-first: Western Electric (done in by acquisition and divestiture), Wurlitzer, and International Harvester.

Over these four event-filled decades, he has learned this one truth: writers either die on the job or outgrow their craft. Most of his peers left the trade twenty years ago. It took him more than half his life to figure out that training is more fun and software testing more lucrative than writing for others.

Then came the biggest opportunity ever—the chance to bid on a project of ecological significance on a national scale. The client was to design and build a system to handle waste and storm water flows, including a sewage treatment plant; this system would help restore a coastal estuary that had been polluted by the effluent of several major cities and industrial centers.

A pair of snowflakes chased each other, flickering through the cone of light emitted by the headlamps of the tow vehicle. Strapped tightly into the 727 awaiting push-off, I wondered what had gone wrong. What could I have done to prevent this situation? Would I ever get a chance to fix it, or was I out of the project altogether? As it turned out, I had been turned out. That, after all, is the fate of the contractor, so the end of my participation did not matter nearly as much as the way it had ended.

First, understand that I am a *writer,* not a communicator.

I know many professionals proud to call themselves technical communicators. The word *technical* places them higher on the pyramid of scribblers—certainly higher than public relaters (which I have been) or greeting card versifiers (which I have not been) and fairly close to those actually published under their own names and earning royalties for their work. I am generally uncomfortable around folks who call themselves business or technical communicators, and I have always opposed applying any adjective to my work. Adjectives can be so limiting.

Business and technical communication seems to be the field beckoning recent crops of English majors. When I was a student, the vision for those not pursuing academic careers was journalism or publishing. The pinnacle of nonacademic career choices, especially for women, was to become a reader for a major publisher: it was real clean employment. And back then, the predecessors to today's business and technical communicators were mostly male—the only factory workers who wore coats and ties with *black* shoes. They either edited house organs or wrote operating manuals.

Back then, all I knew was that to be a writer who wanted to eat regularly if not well, one wrote whatever it was that someone who had the money to pay for it wanted written. Thanks to an appetite that insisted on being satisfied, I labored as a generalist in many different vineyards that have since become highly specialized. Had I restricted my employability by the use of an adjective, I would not have had the opportunity to take the vow never to work for another advertising agency again. I have taken *that* vow at least five times.

My career started in what my first employer called academic public relations. Actually, I wrote college catalogs. But before I got to write the catalogs, I had to write critiques of the existing catalogs. They were always out of style and poorly produced. Armed with my critiques, the sales forces went out to sell admissions packages that were smartly up-to-date and very well produced.

From college catalogs, I progressed into sales training, teaching service-station driveway salesmen how to increase the value of a fill-up by looking under the hood for opportunities to sell TBA (tires, batteries, and accessories). That was when gasoline sold for 21.9 cents a gallon, and driveway salesmen both *pumped* the fuel and *checked* your car. Sales training of that sort led me down the slippery slope to advertising and sales promotion.

My more memorable promotional effort never did see light of day. It was an offer for a commemorative tumbler celebrating the successful flight of *Apollo XIII* free with any fill-up of eight gallons or more. When *Apollo XIII* got into trouble, the promotion changed from successful flight to successful rescue. I sent the carton of samples back to Libby. Had I kept them, I could have retired on their value today.

Eventually, I plummeted to the real depths of writer's hell—direct mail! Direct mail is seldom pretty, but you do know when it's working. I'm still proud that my accident indemnity package ($10,000 coverage for just $2.95 a month) for a mail-order insurance carrier was the control for more than five years. The control is the package to beat, the one that produces more than the others. I still see imitations of that package today, almost thirty years later.

You cannot call such meandering a career; it was more like a wind-up doll moving along a crooked path by careening into walls. I have lost count, but since 1965, I have had at least two dozen different employers and six different episodes of self-employment. I spent twenty of those years as a contractor, writing for hire on a project-by-project basis, as opposed to being an employee in the business. As a contractor, I seldom saw a project through to its conclusion. I was usually involved in another project by the time the video was released, the manual was published, or the presentation was made.

Eventually, this careening led to technical communication—industrial chemicals, electronics, telecommunications, optics, weapons systems, and civil engineering.

Technical communication is where I was first greeted with the word that still makes me crazy: *fluff.* Engineers use *fluff* to describe any text that speaks in active voice directly to an audience, accurately describing a product, a process, or a technology in terms that the audience can understand. Fluff does not follow that linear path in which every step of the process must be covered in the very first sentence; it is more complex communication. Fluff does not strive for emphasis by splitting infinitives with the universal adverb, *effectively.* In short, fluff is what writers do, what audiences appreciate, and what engineers despise . . . well, some engineers.

Most technical communicators seem not to take umbrage at that word the way I do. At least, they don't express it. For the longest time, I didn't let it bother me, either. When I first heard an engineer use it, I thought he was exercising some kind of sick humor, or maybe it was the kind of juvenile putdown feminists call male bantering. Once was an incident, twice a coincidence, but by the third time, I knew this was a trend.

Had I entered the field directly as a technical communicator, I might not have thought twice about fluff. But I had heard the word used elsewhere . . . as a verb. My attitude might have been different were it not for the time my career meandered into direct-mail promotion for a swingers' convention. One doesn't merely dip a toe into the demimonde of alternative lifestyles. Like Pope's Pierian spring, one either drinks deeply or tastes not of pornography and how it is produced. In that industry, the pecking order is specific and measurable. On the lowest level of its social scale kneels the fluffer, the person charged with keeping male performers functionally alert. The service they perform is called *fluffing* or *to fluff.* So this is what that guy thinks of my work! This is what I thought engineers had in mind when they used *fluff* to criticize lively prose. If you didn't already know about this, perhaps learning about it now may change your willingness to accept this word as part of a critique.

So what does all this have to do with my state of mind as I waited for the air traffic controller to allow Captain Carlson and his Pittsburgh-based flight crew to carry me safely back to Orlando and the comfort of home? At this period in my life, I had become the sole writer for an audiovisual production and event-staging company. We produced presentations and business events, and I was contracted out to develop the content for those events.

A regional office of an international civil engineering firm had come to us with a challenge. They were engaged in a bidding war for a metropolitan sewer system project that would run the better part of the decade and pay project management and design fees in excess of $5 million. Although the project was subject to competitive bids, this firm had been the front runner, a virtual shoo-in for the engagement. However, the first round of presentations had shocked them into competitive reality, because their turgid performance had not cut it. The number-two contender had shown a razzle-dazzle multimedia presentation based on John Naisbitt's *Future Trends.* All of a sudden, number one found itself a distant number two. Could we help?

Working closely with the project manager, we helped them develop a solid strategy to present the team's expertise confidently, supported by clear graphics. More important, and far more difficult than rendering their presentation in clear and compelling graphics, was writing a tight presentation that required the presenters to work with a script. It was difficult, yes, but not impossible.

This team had been dead set against a scripted presentation. They were confident that their off-the-cuff comments about a series of deadly graphics seemed more sincere. All they thought they needed for us to do was make the graphics more interesting. We demonstrated that in the time allotted, off-the-cuff would fail to put across all the competitive points necessary to get back in contention. A couple of rehearsals using scripts integrated with graphics showed that not only could they get their points across clearly, but also they could do so with far more confidence than their old presentation style allowed.

The difference came from *working with* scripts, not reading them. Working with a script entails providing a complete document to each presenter: all the words he/she intends to say are down in black and white. The presenters get comfortable with the material by reading it aloud, marking out uncomfortable words and phrases, and inserting others that are more comfortable, more powerful, or more appropriate. As this process moves forward, presenters grow less reliant on the script itself and more confident in the presentation. Eventually, the script evolves into a roadmap, a guide for slide changes, and cues for speakers. It is especially valuable to manage elapsed time where competitive presentations face such constraints. That is what I mean by working with a script.

The team stormed into the final presentation with a lot of ground to make up. Not only did they make up lost ground, but also they forced the process into a two-way playoff, which they subsequently won.

As a result of this win, we had a team of converts on our hands. We had helped them both discover the value of a tightly scripted presentation and see how easy it was to create effective presentations of their own. Those were heady times for us. During the next eighteen months, our firm was assigned to all of the client's problem presentations. We continued to convert off-the-cuff speakers into razzle-dazzle presenters. And our client kept on winning bids with polished regularity; contracts came at a far higher rate than the firm had ever experienced before. Then came the biggest opportunity ever—the chance to bid on a project of ecological significance on a national scale. The client was to design and build a system to handle waste and storm water flows, including a sewage treatment plant; this system would help restore a coastal estuary that had been polluted by the effluent of several major cities and industrial centers.

Teaming with our client, we set out to follow the process that had been so successful in the past. Our producer and graphic artists flew into Washington to meet with the project team located in northern Virginia. After having established presentation strategies and graphic treatments, the producer and artists left the client site, and I moved in for detailed scripting.

The client's project staff included nationally recognized experts in several disciplines with whom I had worked before. Staff professionals from this regional office provided points of contact with the project's funding agency and made up the balance of the team. I threw myself into the work eagerly, and it progressed quickly and smoothly. Those who had not worked with us before meshed smoothly. They found themselves caught up in our enthusiasm. As we clicked steadily toward the deadline, the last issue remaining was to bring the project manager up to speed on the presentation process.

A project of this scope demanded a senior engineer experienced in projects of national significance. The ideal candidate was just coming off a two-year assignment that involved building a national sewer and water treatment system in the Middle East. Unfortunately, he was not available to participate in the planning and proposal development. Our client's senior management was convinced that our approach and the solution we were presenting were so strong that the project manager would catch up quickly to lead the team into another successful presentation. They cut it close, however, because he was scheduled to fly in from Cairo via Rome, arriving in Washington a scant twenty hours before the presentation.

The day before the scheduled presentation, the senior engineer's flight was delayed—bad weather in Rome and here in the States. Our noon meeting with the project manager was put off to 3:00 P.M., then to 4:00. People were getting nervous, worried. Would he make it? Finally word from the airport: his flight was on the ground. We would assemble at 5:30.

In most engagements of this type, my work would have been completed by now, and I would have been out of there. The producer would have been on site to organize the materials with the presenters and to stage the event. But the client was going to stage this one using in-house resources. That was different from our model. Since I had worked so closely with the development team—indeed, I was a key member of that team—my task was to run through the presentation one last time before parceling out the parts to the individual presenters. That, too, was different. The client deferred final confirmation of the presenters to the project manager; this was to be his personal contribution to the process.

So the meeting convened. Around the table sat a collection of the most astute civil engineers and project specialists I had ever worked with. It had been more than a great pleasure for me . . . it was fun. At that end sat a nationally recognized expert on storm water flows. Next to me on the right was one of the world's great wastewater process designers. Across the table sat the U.S. expert in predictive storm modeling. We had all come together as colleagues—friends working together, all except the project manager whom some of us had just now met. All were beaming in anticipation, proud of what we had accomplished together . . . all except the project manager, no doubt fatigued by his journey and increasingly showing signs of fidgeting, like a child at a long church service.

I began. I started by explaining the strategy, showing the graphic treatment, and laying out how the team would stage the presentation. I paused and looked around the table.

"Go ahead, Jerry."

Deep breath, and I launched into the content. I read the first page. Looked up.

"Keep going."

Around the table, smiles of encouragement: comrades all . . . I kept reading.

About halfway through the second page, the project manager stood up, signaled me to stop, turned to his left, to the man who up until then had led the project as his surrogate.

"This is fluff."

Had someone lobbed a hand grenade into that conference room? People scrambled for the door and pulled me with them. Senior staff stayed behind to counsel with the project manager.

The office lobby grew gray with the buzz of agitation. Concern, so thick it felt like smoke, filled the room. Time passed. How much, I don't know. My memory holds no specifics, just the buzz of a project in disarray.

Eventually, I was thanked and excused. I set out to face yet another chaos: Washington National (now Ronald Reagan) Airport on the edge of a winter storm. I somehow managed to catch the last flight out.

The heavy feeling of being crushed came from the suddenness of it all. All that time, all that energy, that huge foundation of success built on months of experience all flushed down the sewer with a single word: *fluff.*

Had I been an employee, I might have been able to help fix it, but I was a contractor. My involvement was no longer desired. The contract had been satisfied regardless of the outcome. No doubt, the personal good will and reputation I had developed over those many months were now being consumed by whatever engine was needed to pull together a project that had been quickly rendered a shambles.

Contractors are hired for their skills, yes, but the impact of those who get hired repeatedly usually comes from the intensity with which they engage in a project. All contracts end eventually. But to have my involvement end by crashing against a wall erected by a puff of fluff . . . this was harshly different than anything I had ever experienced before or since. To this day, it is difficult to recall the details other than the finality with which that man uttered that word.

Today, I am back where I started: in school, working on that master's degree in English I abandoned thirty-five years ago. What I have learned in all this is that a team project cannot succeed without the whole team working together from the start; you cannot add a principal member at the last hour—especially the team leader. What I did *not* learn, and absolutely refuse to learn, is to curb the intensity with which I throw myself into any project. The real fun is still the work, even when all seems to go wrong. You never lose what you learn in working any project. And I do know that the word *fluff* can never be taken lightly.

Reva Rasmussen

Madame Mao in the Midwest

Reva Rasmussen earned a B.S. in nursing from Northwestern University and an M.A. in writing from the University of Minnesota. For five years, she combined these degrees to research and write papers directing medical care, which were published in a software package that is distributed internationally. She also has published news articles, short stories, opinion editorials, and creative nonfiction in various venues, including literary journals, newspapers, and radio. Currently, Reva lives and works in China where she is researching new material.

My first three years as a medical writer were good years. The learning curve was high, but our manager, Cynthia, always encouraged us, and when we had trouble with other departments or staff members, she backed us. The discussions with the doctors were intellectually stimulating, and the docs were generally great to work with. Furthermore, I am a creative writer at heart, and the exactitude required by medical writing helped me hone my techniques for creative writing.

I was one of four writers at the beginning, but the department grew quickly; another four were added at the end of the year. Cynthia did not like managing people, so she left us alone. She gave us the guidelines for our papers and left all the small decisions up to us. We worked with authority and efficiency. It was an ideal arrangement.

Of course, the job was not without its problems. There was an administrative assistant who did our word processing. All of our work had to go through her, and she used an outdated software program that only she possessed. We had our own computers, of course, which we used to write the first draft, but then we'd have to give our papers to her to add the references. At this time, she converted the document into her software; from that point on, she had total control of the document. If the writers needed one word changed, the document had to go back to her to make the change. She did the word processing for eight writers, plus memos and reports for the rest of the staff, so we had to wait a long time for

> *With Jean's encouragement, the editor began to make more changes to our papers and to add mean comments as well. Three people who didn't understand our work were now attacking it.*

her to return our documents. She complained a lot about her workload, but I believe she loved having lots of people waiting for her. It made her extremely important in that office.

Once, I overheard her tell a friend in the office—she had a lot of friends because if you were her friend, you got your work back faster—that she regretted not having gone to college. She knew she was as smart as everyone she was working with, but she just didn't have the piece of paper to prove it. The woman did not have much education, but she had power. She had doctors and nurses and writers trying to please her so they'd get their work back from her when they needed it.

When a document finally came back, it was full of her edits. They included simple things, like spelling out abbreviations or correcting spelling, but she also changed the wording. Sometimes her edits were acceptable, sometimes not, but the big problem for us was that she made a lot of typing mistakes in the other parts of the document. So, I'd get my paper back and have to mark her errors and give it back to her. She was never able to correct all her errors because she'd get distracted by the editing she was doing.

We complained, but Cynthia defended her, saying her edits were good. I think the real problem was that Cynthia didn't want her to quit because the unemployment rate was so low that it was hard to find a good administrative assistant or even a bad one, especially someone who could use the software to which this person had converted all of our documents! She had been asked by the information services department to change her software, but she always convinced Cynthia that it wasn't necessary. We sometimes wondered if she clung to the outdated software in order to protect her job.

We didn't have an editor when I started, and I helped Cynthia choose her from the applicants. Our company wasn't offering a good salary, and the job requirements were high. Applicants who were qualified for the job wouldn't take the low pay. Cynthia and I talked openly about this at the time. She wasn't happy with the editor either, but she was the best we could get. So she was hired.

The new editor presented new problems. She had degrees in biology and zoology, but nothing in the health sciences or English. She often did not understand the medical papers that we were writing. We could have worked with her if she had been willing to listen to us; we all had backgrounds in health care and writing, but she was a very defensive person.

Back then, Cynthia and I were good friends. We often discussed the problems the writers had working with the editor. Cynthia agreed with the writers then and helped us in our disputes. At that time, she gave writers final say on the edits.

After I had been at the job for three years, the writing staff increased to fourteen. One of the new writers, Jean, said she had a Ph.D. in public health. Oddly, her previous job had been managing a restaurant for five years. Before that, she had worked for the state health department in a low-level position that didn't require a doctorate. It was a weird work history, and the writers were surprised when she was quickly promoted to be our supervisor on the grounds that she had a Ph.D. Cynthia gave her free rein and removed herself completely from us.

We noticed the new supervisor was good at making lists and timetables for our work, but she didn't seem to be good at other things. She often complained that our papers were unclear and wanted more explanations of medical terms. When we protested that we were writing for doctors who would understand the terms, she snapped, "I'm a doctor and I don't understand!" There seemed to be a lot of questions that we shouldn't ask Jean. When we asked her medical research questions, her answers were evasive. When Linda, one of the writers, asked her what she had written her thesis on, Jean answered, "It was something you wouldn't be interested in." Linda recalled that Jean's résumé, which had circulated among us before she was hired, said she had earned her doctorate at the state university. Linda did a search of theses in their library but was unable to find anything in Jean's name.

A rumor started among the writers: perhaps Jean did not have a Ph.D. We discussed this among ourselves, but no one had the courage to go to Cynthia with our suspicions. Things got crazy.

With Jean's encouragement, the editor began to make more changes to our papers and to add mean comments as well. Three people who didn't understand our work were now attacking it.

Although my friendship with Cynthia had grown distant, I still trusted her. I knew she was naive about people, and perhaps she didn't understand what was going on; after all she talked to no one anymore but Jean. I arranged to have breakfast with her one weekend.

We had a nice meal; we both expressed regret that so much time had passed since our last social outing together. Cynthia was working on her own doctorate in public health, and she updated me on the progress of her thesis. Then we moved into our personal lives. We both had aged parents who lived alone, and we worried about them. I felt relaxed, and the wonderful feeling of closeness we had enjoyed for three years was restored. So I quietly brought the conversation to the subject that deeply troubled me.

"We writers aren't very happy at work." Cynthia looked surprised. How could she not know? I wondered. Was her reaction honest? I continued. "We all love our work, but we don't feel confident about Jean's leadership." Cynthia was again surprised.

"Why not?"

"When we ask her research questions, she doesn't understand what we're talking about. She gives us answers that don't make sense. When we ask her what she means, she changes the subject."

She sat back and said slowly, "Ye-e-es, I've noticed that."

"It's surprising too that she never talks about her work in public health. You know, like what she did for the state department and her thesis. Doctoral work takes up so much of one's life, it requires so many hard sacrifices, that people always want to discuss it. Has she discussed it with you?"

Cynthia shook her head no. "I asked her once, but she had to make a phone call, something urgent. There's always so much to do."

"That's what she does to us! Once or twice, okay, but always? And when you're in the same field? She won't talk to anyone about her thesis or graduate school. She says it was so long ago, it doesn't matter. And she won't talk about what she did at the state health department. As far as we can figure out, this is the first job she's had that required a doctorate. It just doesn't add up." I jumped ahead with my message. "We don't believe Jean has a Ph.D."

Cynthia caught her breath, stopped breathing altogether for a moment. Her voice, when it came, was not steady.

"I gave her an article on methodology, and she gave it back to me saying it was very interesting, but I don't believe she even read it."

"You see?! Of course, it's circumstantial evidence. We're just guessing; we could be wrong, but things don't add up." Cynthia looked stricken, but she nodded in agreement.

"I wondered why she never read the journals I gave her."

"They were over her head," I said flatly, then regretting my bluntness, I added, "or maybe she's just rusty. After all, she was stuck in that restaurant business five years before she came to work with us. Imagine how displaced she must have felt. A Ph.D., and there she is, managing waitresses. I wonder how she ended up there." I started to laugh, and Cynthia joined me in a high, hysterical pitch. I could feel her tension. I stopped laughing and looked closely at her. It was hard to read her expression, but there was something of shame in it.

"Well, it wouldn't be your fault if she doesn't have a Ph.D." I wanted to reassure her, to smooth over the problem. "The human resources department is supposed to check people's degrees, but they never checked mine. They're the ones who should check these things."

Cynthia's eyes were watering and she looked pale. "I'll check this out on Monday."

On Monday at 5:00 P.M., Cynthia came to my cubicle. She didn't look at me, she didn't face me; she stood to the side and looked out the window. She only said, "I put the ball into motion." I waited for her to say more, but she was silent.

"We'll hope for the best," I answered. She walked away.

Tuesday was the last day we saw Jean. Friday, Cynthia sent an e-mail to all the staff. Jean had resigned. No explanation of why. I walked to Cynthia's office. She looked up at me from her chair, then back at the papers in front of her.

I stood awkwardly in the doorway, then stepped in and closed the door.

"What did Jean tell you?"

"She claimed she had a doctorate, but she couldn't produce a copy of either a Ph.D. or a master's degree." Cynthia continued to look at her desktop. "We gave her a choice: tell us where to find your degrees or resign. Well, she's gone now."

I breathed deeply and smiled. "This used to be a good job. Now we can get back to where we were before." Cynthia still did not look at me.

After that, we expected things to return to normal, but they got worse. The editor's edits became savage. She often required us to rewrite. Her requests didn't make sense, but every time one of us tried to discuss it with Cynthia, she became angry and defended the editor. The administrative assistant's errors increased as she also edited our documents more extensively. To make things worse, Cynthia decided working drafts of our documents had to be free of all errors before they were sent for doctors' comments. When we protested that this was impossible because of the extensive edits and errors by the editor and the administrative assistant, Cynthia angrily told us if we managed our work better, we wouldn't have a problem.

Cynthia had lost her self-confidence. The writers, the competent people, were a threat to her. She knew that we knew she'd made a mistake, and she was afraid we'd find more of her mistakes. Like hiring the editor. We were dangerous, and we needed to be punished. So she let everyone attack our work.

I began to lose hope that the writers would ever again be treated fairly and with respect. Not only did Cynthia have her personal limitations as a manager, but also it's a marketplace of supply and demand for workers as well as products. There were not enough skilled office people at that time, and the editor's position demanded a high level of ability in both medical language and writing that was hard to find. So Cynthia needed to keep them. Unfortunately, Cynthia didn't have trouble finding a replacement when a writer quit, so she didn't have to listen to our complaints. It was free enterprise, pure supply and demand.

Finally, I realized I shouldn't and didn't need to keep a bad job; I have a good education, good skills, and lots of opportunities.

When I gave up hope of the job getting better, I decided to go to China. Crazily, it was when I started to read about China that I had a framework for understanding what I'd gone through. We'd had a micro–Cultural Revolution. The writers were intellectuals whom Cynthia saw as a threat to her authority. Cynthia became Madame Mao; she surrounded herself with incompetent people and gave them power. Step by step our authority to control what we were writing was taken away from us.

What happened was sad, but not the end of the road for me. I'd wanted to see China for years, and by leaving that job, I was free to explore new possibilities. In fact, writers are very free; with the Internet and a fax machine, I can write as easily from China as from my hometown in the Midwest.

Lenore Weiss

Stranger in Paradigm

I became a technical writer by accident. As a youngster, I attended a high school that had a rigorous science curriculum, but I always wanted to be a writer. I spent my youth working for community-based newspapers and interviewing members of the Vietnam Veterans Against the War (VVAW). I visited the Veterans Administration hospital and talked with paraplegics whose bodies and lives had been blown apart on the battlefield. It was an education, and the best kind.

After the peace and the women's movements had subsided, I found myself with an unemployment check to bolster my slim financial resources. I decided to earn a master's degree in creative writing at San Francisco State University. I realized that a degree wasn't going to help me make a living. But with the cheery recklessness of youth, I decided to go for it. True to my own prediction, after I graduated I found myself a Kelly person at different temp jobs. I kept writing poetry, plays, and stories. My day job, however, was giving me serious grief. Office work, even if it was at the University of California at Berkeley, wasn't fulfilling. So I spent my off hours sitting with a cup of coffee, scanning the want ads.

One day I uncovered a curious proposition for a technical writer. I didn't know what a technical writer was, but I did recognize the writer part of the description. I thought I could handle the job and applied. My first assignment as a technical writer was to develop a proposal for a wastewater management system. I made up for what I lacked in skill by working long hours to get it right. My next assignment was working on a computer application for bank tellers. I didn't realize it then, but I was hooked.

Not only did I buy my first 64K computer, but also I became active in a local computer users group where I edited the chapter newsletter for BAMDUG (Bay Area Micro Decision User Group). Here I discovered what it meant to talk with another person in real time across a modem line, known simply today as chat. I could sit in my own living room at my own desk and converse with someone else in Florida or the Martinique Islands or anywhere else in the world to exchange perceptions and experience. I was amazed. Actually astounded. I immediately understood that what lay at the heart of the computer revolution was communication between people. At the same time, I also discovered The Source, a predecessor to groups like AOL (America Online), and later joined the WELL (Whole Earth Electronic Link). In those days the WELL was housed in a white wood building in Sausalito, sandwiched between trees and the pier. I drove there a few times to speak with the resident UNIX guru. But mostly I remained within the bowels of different data-processing organizations as they expanded and contracted.

Author's Preface: Bookmark to the Information Age

The printed word is sharing space with the electronic one. You only have to watch television or surf the Internet to acknowledge that electronic presence. Today there are technical writers who translate technology to an ever-widening audience. "Stranger in Paradigm" is part of that translation.

It was the time of the corporate reorganization when people disappeared overnight, escorted to the lobby by security guards and carrying a cardboard box. From my humble cubicle, I unknowingly served as a witness to some of the less glamorous moments of the information age. Others have chronicled the entrepreneurial greats and visionaries who shaped its formation and made big money. "Stranger in Paradigm" is more a view from the trenches, what it felt like to be there.

This collage is my personal response to those times, filtered as a technical and creative writer, a journalist, a mother, and as a former sixties activist struggling with an old working order. But the central kernel of the piece is my search to find a new way to use language that is more expressive of our interactive age.

Modules. Mix and match. Pastiche. Collage. I continue to ask myself, How do we create meaning and context from these news byte-sized pieces and then reassemble them in different ways? My experience as a mother has helped inform this process. Jumping from one thing to the next without advance warning has become a way of life. It's this sense of constant discontinuity within a continuum that I've tried to re-create.

Part I

The Prelude

I sat at my desk for four months, waiting to be transferred to another department, and because none of my superiors exactly knew when that was going to happen and since I no longer belonged to them as a real resource who could be counted upon as the equivalent of a full-time employee, the best they could do was ignore me. Of course, every so often they requested I format a letter or design a brochure, drawing me into the world of doing verbal things, but it happened so infrequently that these occasions seemed mere anomalies in my otherwise unscheduled time. Truthfully, no one gave a good triplicate form what I did during the day, and this, more than the fact that I had no work to do, came close to corrupting my spirit. Instead, I decided to become a desk. Not a real desk, but a piece of furniture, quiet, with drawers I could retreat into,

> *Why does it become the norm for everything not to make sense (because sense is composed of a myriad of potholes and overlapping roots all competing for water)? Research: root systems, what are they like? I'm posting this note for you, babe.*

where no one could give me the latest gossip about which department was being dismembered or who was on the cut list. I counted the number of pushpins residing in my stationery tray. I arranged my paper clips so that they all faced in the same direction. Sometimes I worked on my computer, but I'd been through the tutorials so many times before that I chose to turn on the screen saver and remain inside my desk. Comparisons with the womb are obvious, but it was the construction of the drawers that really fascinated me. I've always been a person who likes to know how things are made.

Rabbet joints are common enough, but it's the fit between two planes of wood and the care taken in its construction that are crucial to the futurity of a piece—for example, if the wood was originally sanded with several grades of paper and if the glue was allowed to set. These things take time; you can't hurry them. Slowly, I begin to see that the drawers of my desk are of medium quality, the wood a kind of composition board with walnut veneer. I want to endow it with more dignity, even though the handle of one has been totally lost in its last reorganization, and the handle of another is coming loose, its screw revealing spirals of pink paint underneath. Any contribution I can make has to be made from the inside. Cosmetically, the desk is a wreck.

Each day I review the progress I make in organizing the insides, move the paper clips closer to the front of the drawer, and decide that I really don't need to save copies of all my time sheets as long as I have copies of my paycheck stubs. This frees up more room. The pushpins stick out like miniature daggers, and I know what to do: retrieve my time slips from the trash and tear them into confetti-size pieces, throw them at whomever cares to listen. I am bestowing membership on a new order where our days are not divided into REG hours, a piece of paper which has no meaning other than to give our time the exact names of the week.

I gather parts of the grid into my hands and release them over my head. They settle around my waist and orbit, a meteor shower. This makes it difficult for me to sit in my chair. Hornets whirl around my torso, a Van Allen belt. I reach for my purse so I can grab a cup of coffee. I barely fit inside the elevator. Everyone is looking at me. I drink my coffee decaf with low-fat milk. I arrive back to the cubicle after my coffee break, and I see that someone has swept up the time sheets from the rug to hide my indiscretion. Who?

Alarm Clock

Today I am an entrepreneur and do whatever I want. I innovate my own meaning. I am building my own business and encourage people to call me at work. My voice mail answers messages perfectly. I am building my own business. I am being an entrepreneur and listening to my own inner voice. Somewhere there is bliss. I am lost, left without instruction, trying to find you at the amusement park, and all I see is the tattooed man and the fat lady, have this really sick feeling in my stomach knowing that I'm going to dissolve the way cotton candy does in my cousin's mouth. I want to be an entrepreneur, but I know we came to the park together and that we're supposed to go home.

First Cup of Coffee

I need to pick a server conveniently located at my nearest node, a place where I can dip into the well of the universe and taste the water, and wonder how so many Chinese women writing from inside the civil service system in the eighteenth century were shunted from one province to the next because their husbands fell out of favor with a certain official. Guess what my days are like.

Traffic

Radio countdown to destruction liquid cool underwear duck my head in traffic waiting for the red light too long while a driver meanders between lanes, doesn't he know, I swerve, use the blinker, something in the car's back trunk thuds, doesn't he know; there's a run down my tights, a check I have to write, a phone call to make, and who knows what are we going to eat for dinner? Thank God. I did take this month's new parking stub to hang on the dashboard. I can't fit into small spaces. That's why I dumped ice water on the construction worker's lap in the Korean restaurant.

Turning on the Computer

For a week I didn't sleep, slowly admitting the truth to myself about a period that would never come, doing what I did when I was a nervous girl wondering if this month I had really gone too far, crunching a ball of toilet paper in my hand and rocking the top of my uterus, hoping to strike it rich.

I was hungry immediately, smelling out thick barley soup, shiny with rafts of white mushrooms; at every street corner, I wanted to eat. Why am I having a baby again now? An unanswered question, a need, an urge. This morning when we made love with the sunlight filtered through the white muslin curtain, my nipples were as sensitive as two joysticks to the pressure of your touch. And I can remember, disengaging for any instant from the circular motion of my hips to introduce you to the baby. What I mostly resent is being pulled back to an earlier imagery.

Listening to Voice Messages

There were 256 shades of smog in the sky this morning, radiating outward from the bridge, a halo that encased me on the toll plaza. Sunnyvale doesn't even desktop-publish their employee newsletter and they're at the center of the Silicon Universe Valley. Plus they don't have a whole lot of use for PG&E either, sending out their own utility bills and generating electricity while the storekeeper's daughter's taking care. The calendar's looking dated. We'll transmogrify the *California Living* publication into a community awareness piece and not even talk to the people who've been doing it for the last six years. That doesn't make sense. Why does it become the norm for everything not to make sense (because sense is composed of a myriad of potholes and overlapping roots all competing for water)? Research: root systems, what are they like? I'm posting this note for you, babe.

Time to Work Out: The Warmup

Jobs do not exist for people to be creative or even to work with other people who are about a similar task. Jobs serve as a power base for the head in command, as a reserve to throw down the hole of the next budget deficit. The notion that people work at their desks for seven hours is a myth. People make appearances at their desks and find innovative ways of demonstrating that they are doing a job by inventing their own projects, which serve as the real way to make time pass from Monday to Monday. Bills are paid, letters written, teachers contacted, appointments and reservations are made for a long weekend. The real work is everything that surrounds the job. Work has nothing to do with what we traditionally think of as work, a concept that harkens back to man as hunter and woman as matriarch. Work is self-generation, a daffodil of light that grows incandescent around my stem. On the jogging track, I pretend I'm running on a three-mile stint past redwoods and eucalyptus groves near the Horse Arena. On the way back, the trees sing to me. I find twig people; one lizard, one dancer.

Second Cup of Coffee

Outside my window there are other windows with reflections of buildings twirled inside a ribbon haze of light. On the sidewalk, a parking meter wears a condom so that it can't receive quarters. The latte is creamy, almost sweet. It's difficult for me to stay in one place. Yet everyone sits by some kind of a window. Even in my own house, the cats chase each other back and forth and jump on my headboard to hoard the sill. And what do they see? Birds. I like keeping the latte in my mouth until it turns into sea foam. The desktop is not a metaphor. It's an organizational principle.

Interview

I visited Oracle where the towers are marble green with three different gourmet restaurants, an employee gym, cleaning service, and photo service, all on the premises. Employees can order cubicle-size or office-size flowers. I parked the wrong way in a stack of waffles that was really a parking lot, in the bottom-most waffle, feet getting stuck in the syrup, which is what happens sometimes when you ask for directions, "Just down the staircase and to your right." Glop, glop. Scattered car parts between parking spaces, things I never knew could fit inside an engine, not that I'm such an authority, but I have spent my share of time waiting for my tires to be balanced.

Finally, I found the elevator. I walked across a glass bridge, saw a lake surrounded by willow trees, and pushed the first-floor button. Oracle's lobby continued for endless versions. I understood. This was not a place of work. No thank you, Mabel, child prodigy and MIT graduate.

In the Elevator

There's a man with a cellular phone, his personal link to another consciousness. The ends of his beard are braided, stayed at the ends with orange and green beads.

"What's wrong with you, man?" he asks between the second and third floor. "You mean that?" He repeats this again with more emphasis. *"That?"* Several of us look at each other. He has already created an intimacy. "What do you mean I should do something?" He drops the phone from his mouth for an instant and stretches his fingers toward the control panel. Only the eleventh and fourteenth floors are lit, but the elevator stops at the fourth floor, and more people get in. "You see now what you made me do?" he says. "It is not a question of blame," he says, standing next to a suited man whose eyebrows are beginning to wrinkle across his forehead, looking over to one of us hopefully. "As a matter of fact," he says, pulling the edges of his caftan from the clutches of the elevator door, "it's getting crowded." We all laugh, "Hea, hea, hea." He gets off on the next floor, still talking in his ear.

Weekend

I take your leave and brush my daughter's hair with saliva, one strand at a time.

You were born in an aftershock of an earthquake, people still buried in the rubble of the Cypress Overpass. No one believed in having babies that night.

I watched you breathe for several hours, your lips blistered with milk. Twice you opened your eyes and smiled at me. There were no words, just milk.

Under the blue tensor light, the divining rod of your mouth seeks out my nipple.

An umbilical cord is a silk rope wet with the dew of birth, a blue horizon unwinding to the very ends of you, a tassel.

The staircase of your fingers spiral to your nose, a rope lattice for the air to climb.

If I had to choose, it's your feet I love the best. The way you pushed yourself from me like a swimmer shoving off from the edge.

Phone calls to make, rooms to vacuum, the wash is in the dryer; we listen to music, your five fingers pressed to my hand.

You recognize the plaid of my bathrobe. I'm the magician who turns off and on the light bulb while you play kiss with the sun.

Dead jellyfish wash up on the beach. We find the snaky remnants of sandworm poop, coiled basket of cobras. A shark float bounces along the sand with the kids running after it. You collect wads of seaweed to make salad on the beach. Barbara shows me the six rolls of toilet paper in her new bathroom and jokes that it's time to go back to Costco to buy a new package of twenty-four. Jiggedy-jig.

He would climb a mountain to punch an echo. He could throw a lamb chop past a wolf. Are we getting on the edge of the diving board here? All my squirrels are running up your tree. I crush my body with my mind until the shaking is squeezed out of it. Mabel has bought a cardboard condo in the valley and left me with one word of advice: don't let a witch woman take care of your baby because she will feed her chewed rice and uncooked beans and change the baby into her own liquids and what's running through the baby's veins will draw her away from me. I sit near the telephone and hate everyone who returns my phone call. These are the ones whom Mabel warned me against.

Part II

The Prelude

I am going to work tomorrow, and my daughter will be in child care. My husband says it's all right. He is trying to comfort me. "Do you want to stay home?" We both know that I can't, but he says I can do whatever I want. I don't know how to give myself that courage.

One day I will give myself a transfusion from the snow pack that turns into water each spring and hangs next to my earrings and hairbrush. It's so strange that I have earrings and a hairbrush because once I was a feral child in the abandoned lot who only came upstairs for dinner, and really then only for bread, a thick crust. But there is no witch taking care of my baby. I keep telling myself that is a child's story, Mabel's story who thinks she can do everything.

"Do you want to get a nanny?" he asks. This is getting ridiculous because we both know that we cannot afford a nanny. But I know he is trying to be kind, and takes my hand over the dinner plate. My answer is Yes/No, a branching pattern that lets in no sunlight.

Yes, because who wouldn't want to in my situation, and No because I don't want my Persephone bathed in fire and made immortal. I long for my mother to be on this earth because then she would know. But would she really, from a generation of women who were involuntarily joined at the hip to their children? A doctor commemorates the death of a Siamese twin who was about to reach her first birthday by saying she knew how to laugh and had a great smile. I read a newspaper article about the emergency room around the corner from where we live, where doctors deliver guns wrapped in plastic bags from the vaginas of fifteen-year-old girls. I try to tell you, I am afraid of the violence that freezes our hearts. Isn't leaving our daughter so young an act of violence? I have to get up from the table because the baby's crying.

Alarm Clock

Time is the fourth dimension, and I am living in fiscal year-end budget time, an animal who hibernates during the spring. I burrow deep inside my life where there are no budget hearings and let others go on retreat to compose long agendas so they may sculpt their time according to bullet points. A point may be a wave, and each issue takes an hour to discuss, more like two. Council goes around the table wearing plumed hats and T-shirts, thumbing through their packets for a paper to explain everything.

A new ice-skating rink is going to be built on San Pablo Avenue, and legislators look inside their briefcases for pom-poms. I am studying field theory inside my burrow. A worm shares my apartment. She eats and excretes soil, which makes a lovely mound right next to my desk, a place for my feet. Sometimes I find items that she cannot use, a gold ring with a turquoise stone. Today there's a new flow chart being circulated beneath the door.

"This is which way we make our decisions so early in the morning," sings the daily newspaper. I'm not sure which way is up. Does it matter? I decide that I'm going to be a worm also, a bosom buddy while the air smells of cinnamon and spice and the white plum blossoms float down from the sky. For every action, she does something else. That's what's so maddening. I can never know how to introduce my agenda when the general symmetry is all over the place. Cronus X is one of the nicer assholes around here. Maybe he'll listen. I curl myself into another dimension and compress my time into an envelope the size of a paper clip. I wonder where I've been all these years. I have a twin that exists between the hours of 9:00 and 10:30 P.M., maybe until 11:00 P.M., if I'm lucky. On the morning train I see everyone folded into an accordion. I hear music, but there's only phone messages. Time and money warp me out of shape. Woman Travels through Wormhole to Escape Creditors.

First Cup of Coffee

I started at Primos, next to the Bart station, a small regular, please. Several counter people were in their early twenties, pierced noses and tongues, and Victor whom I liked the most from Sarajevo because he made a great latte after my stomach forced me to let up on the caffeine. Primos also runs a cable cart in the lobby of the APL building because ever since the restaurant closed, people have to go somewhere.

Then there was my brief flirtation with Starbucks, Peet's Oregon nemesis, which is strategically located opposite the Federal Building. But now I buy my coffee from Mon Ami, a mom-and-pop store in City Center.

For less than a buck fifty a day, I've learned to discriminate between different kinds of coffee beans, decaf and regular, foreign and domestic, to taste a latte that hasn't been ruined by a sprinkling of cocoa. That's why I decided on taking a cup along for the ride into the wormhole. Clearly I couldn't rely upon finding a coffee stand anywhere close. Plus, the coffee cup signifies my ability to make real choices, something like my own personal time capsule. What else is left, I asked myself, although I'm still thinking about showing up at a City Council meeting and letting 'er rip.

Soon as I parked the car and got close to the office, I stopped at Mon Ami's for my regular cup. Pop rang up the cash register, and Mom turned on the machine, milking the espresso spigot with a clean damp cloth. Two minutes later, I had my coffee and was about to catch the elevator upstairs. I turned to get a plastic cover, and for the first time I saw it: a whirlpool the size of a thimble kicking up the foam in my latte. I reached for one of those white plastic spoons from the counter and tried to stir down the ripple. I got pulled in.

Traffic

I heard a voice cry, "Call 911!" but I am recycled through the other end of the cup until I bottom out. I don't have a clue where I am, on some kind of a grid that reminds me of my attic, just the wooden beams and nothing else in between except for the insulation. I realize that the spongy stuff is like the foam from my coffee, but on a grander scale. I am going to be late for work.

Turning on the Computer

Deep inside I am scared with my adrenaline pumping. Each time I take a step, I can see myself go through the permutations of lifting up my foot and then putting it back down again. I see a thousand reiterations of my geometry on a back-lit screen. Everything I do is sampled back and forth until I have no idea if I am moving. Plus, I think about my kids and my husband and how I'll never see them again. It's only morning, and already my day has become such a mess.

Listening to Voice Messages

She is in the second realm and comes to me on powder-puff feet, smelling of soap and talcum and yeast that's discovered warm milk. She is exactly the size she wants to be after years of dieting and I congratulate her on looking so well. She smiles and flashes the wires of her bridge hooked over her teeth and points to one of the beams for me to sit down. The beam automatically molds itself to my contours and creates a Louis XIV chair. I rest my elbows on the flowered upholstery. I can't believe this is happening.

"How are you doing?" she asks, and I say a few things about how rough the morning has been. She agrees that some days are like that and gathers me to her side with an arm that sweeps out from the insulation. Her hair is black and her lips deep red. She explains that many of the slaves didn't want to leave, and I ask what she's talking about. They did not want to leave Egypt, she says, regarding me as if I might be slow, even though she's my own mother. She explains that she's assimilating her reality also.

The words move out of my mouth at different sampling rates and I can't understand what I'm saying. She bends down and holds my palms inside her own. What I'm trying to tell you is that we couldn't talk about the destruction, so we moved as far away from it as we could. She looks at me more kindly now. Learn how to select things, she says. She does a slow molecule fade over the insulation and then it's back to me.

As I walk, I find an occasional penny strewn along the beams. The pennies have been minted in years that don't exist yet. I slip them into my pocket for my son, if I get back, that is. Only then does the immensity of my situation hit me: I am a flounder in the universe swimming with my eye to the floor. I throw a penny into the insulation and make a wish. I watch it drop.

Then I see him across the beams, the man with the cellular phone from the elevator, and he asks me how my day's going. I say, "Sheesh."

"What a day, is right," he says. "There were hardly any coupons from the newspaper: just one for an ice-cream sauce with taste-alike pieces and chocolate sprinkles. And some blue-looking stuff to clean your carpet, but I don't have a carpet at home, just the linoleum tile from the other tenant, and a coupon for some of that fancy printed toilet paper. I tore that out, not because I really wanted it; the plain kind is good enough for me, but because I was so upset about not getting a morning appointment with the mayor."

"Forget it," he said. "Bonanza days are coming soon, and you'll catch me in line down by 41st and MacArthur waiting to buy a jar of artichoke hearts; I enjoy those in my salad with a few sliced green beans and a red bell pepper. Now that's what I call living.

"I'm going to do it right, that's what I say. Every week I look at the newspaper, and separate my coupons into three piles. I've got a free pile, a save pile, like when they say 25 percent off, or fifty cents off, and an everything else pile, like a hummingbird feeder or commemorative plate but you have to send away real money for those. I'm wise to them. And then I take each one of my piles and organize it alphabetically, like *C* for all the cereal coupons: All-Bran, Cheerios, Froot Loops; you know what I mean, and whenever I push my shopping cart to the Safeway I carry my coupons in these same piles with a rubber band around each one. I move slowly because saving money is something you need to take your time with if you're going to do it right."

The Meeting

Everything spins in a circle. The sky, the earth, the sun, and moon, people, and the waltz. I hear music but there's no one there. Then I come across an egg balancing on one of the beams. It's sitting in a pile of what looks like salt. I pick the egg up and realize that it's actually bone china with "Made in England" stamped on the bottom. The two halves are carefully latched together and when I raise the top, there are hundreds of candles burning inside the bottom, each one a different size.

The candle within the egg is a life waiting to be nourished. I announce that I'm not cooking dinner tonight and that I plan to see a movie. Nature exhausts the energy of the parent in the launching of the child. But I recognize the possibilities and float across the insulation to a house near the beach.

Still ringing in my ears, the loyalty to products, why change now. Loyalty, standing by and saying nothing, going to the same monthly meeting for six months around the same table and getting work done badly, being in the basement together and bringing up a system that doesn't work because no resources are given to supporting it.

Loyalties are made in rooms and in coffee shops and in the elevator. Loyalties are made in the parking lot following Council meetings and at farewell luncheons. Loyalties are the number of coats you paint the wall and knowing together where the dirt comes from. Loyalty is not always trust but can be the unspoken word between those who have decided to temporarily put aside their own ambitions in the quest for power, an agreement to tell one truth. These are called alliances.

Time to Work Out: Optimum Heart Rate

We never make love anymore. Of course, we might stumble across each other at the end of the evening news before we collapse from exhaustion with a quick kiss to signify goodnight. I'm not talking about that sort of thing. I'm talking about what we called making love twenty years ago, when our bodies were sleek and lean and itched with undiscovered corners that sang back when we touched them. I'm talking about burning incense and taper candles throughout the house with tins of smoked oysters and a good white wine, going at it for hours with an old movie here and there, several changes of clothing draped on the backs of chairs throughout the apartment, and if you forgive me for saying so, a great big fat joint of Wowie Maui rolled up on the table, all the better to admire Artie from several angles and arched in different kinds of light. I was no tub of lard then either, back when we were young people, waiting for the refrigerator to spit out another round of ice cubes for our pleasure; we indulged each other for hours, for days, that is, before parenthood staked a claim on us. I can't remember the last time Artie and I enjoyed an entire weekend together away from the two kids. So can you blame me for laughing at him in his face when he suggested that we hire a baby-sitter and send the kids away for an entire weekend?

Second Cup of Coffee

Some weekends my daughter naps and I celebrate the gift of time, a few hours when I go down to the basement to an area called my room and gasp, like a fish who's miraculously been thrown back into water. In a few hours I'm caught by another bevy of activity, metamorphosing between my job as a civil service employee before I return home to pass through multiple curtains. Now I hear the electric chirp of my neighbor's car alarm. Where are the real birds this morning? I get into my own car and drop off my daughter at child care. It's in the car where we have our most intimate discussions.

Interview

There's the man who stands with a cardboard sign, the letters drawn in black marker, at the intersection of MacArthur Boulevard and Lakeshore Avenue where a red light suspends people for a moment in the traffic. The sign asks for work and money to buy food. The man watches for a rustle of hands beneath a glove compartment, a slit of window from the driver's seat. The drivers on the outside lane keep an eye on the drivers on the inside lane nearest the curb.

I, too, am a spectator, participating in this organization called the City that's composed of departments, committees, groups, different layers of insulation, some of us located at the center of the solar system or on exterior orbits where barely a memo ever filters through, sometimes at a desk located in between, each of us each here for the duration of our work careers, or until the next layoff, or until the next election, running into each other inside the elevator on our way up and down for meetings, coffee, more meetings, where we exchange hellos and mutual support for the day that lies ahead; rubbing up against each other in a constant friction that is politics, a process that shapes the entire body into our collective image. And when one of us retires, another arrives to take our place.

But it's true, Prince Harry, my other passenger; attitude can make difference. The energy you bring to the job, your own particular chemistry when you face out and fan the organization, your work style, your way of connecting, of adhering to principles. For in the final analysis, we choose to work for a bureaucracy where we feel the least personally compromised, where our essential being can still remain intact. Yet even so, as we keep contributing to our pensions and to our deferred compensation plan, year after year after year, the price of that allegiance becomes apparent as we finally become what we first must hate.

In the Elevator

I'm a package of popcorn burning in the microwave oven. I accidentally drove into a crowd of seagulls on Clay Street and killed one. I place the car keys back into the bottom of my purse where I discover a piece of my daughter's old baby blanket, a shred of a blanket really; loved so much, knotted in her sweat, her cuddles, whatever batting is left, filthy and wiped on the kitchen floor. I bought the blanket originally when I was about eight months pregnant, barely able to walk through the downtown Chinatown festival, a yearly September event sponsored by the local Chinese Chamber of Commerce with music, crafts, and tons of food steaming in the street and smelling directly of heaven, outdoor stalls, little roll-ups of things. The blanket was handmade by a Laotian woman. I'm still trying to think of how it originally looked, a quilted square in the middle surrounded by a field of flowers on a light purple background. Over the years, the blanket began to unravel after so many washings and hugs, the stitching began to come loose even after I took out my own needle to hold it together.

Weekend

Pray standing in line at the cash register for meals that perfume kitchens with the anticipation of sitting down. Pray for the music on the car radio that makes the traffic flow smoothly. Pray for parking spots that appear in unknown places.

Dale L. Sullivan

Afterword

Talking to the People Who Know

Highway 41 south of L'Anse and 141 south of the junction were free of snow. I could now step down on the accelerator of my Geo Metro and listen to the hum of its three pistons as I zipped across the Upper Peninsula of Michigan. There was hardly ever any traffic on this stretch of road, but you do have to keep your eyes open: one of the five cars you were likely to meet might be a state patrol, and there might still be a few deer venturing out despite the high snowbanks and the thinning of the population during deer season a couple months back. It was a cold January Tuesday, and I was headed for Milwaukee to speak to the chapter meeting of the Society for Technical Communication. Had to be there and ready to go by 5:00, Central time. That gave me about seven or eight hours.

Several months earlier, back in August, Mollye, the president of the Milwaukee chapter, had contacted me, asking if I would be willing to come talk about the new book Jerry Savage and I were putting together. It was a collection of stories written by technical communicators, and the Milwaukee chapter had passed along our "call for stories" to its members. Now, she said, they'd like to hear about the book.

The impending talk haunted me for a couple months after the arrangements were made, because I hadn't had time to work through the stories with as much attention as my partner had. During Christmas break, however, I edited all but one of the stories and wrote a story of my own to finish the volume. Just this past weekend, I edited stories on my laptop while nominally paying attention to my daughter's volleyball tournament. Yesterday, I cranked out seven or eight transparency slides and promised myself that the outline for the talk would emerge during my drive to Milwaukee today. So on the way down to Milwaukee I was preoccupied: What should I say? In what order? Are these slides (the title of the book, the table of contents, some quotes from narrative theorists, a bulleted list of "what stories do" phrases) going to be of any use?

I slowed, wondering if I would recognize the Amasa cutoff so that I could cut across to Iron River and go south to pick up Highway 32 in Wisconsin. Spotting the turn, I slowed, downshifted into third, and turned onto the snowpacked road. Looks okay: at least the plows have been through this morning. After a stop at Mac's for breakfast biscuits in Iron River, I took off, straight south, past the Ski Brule road and on into Wisconsin. Since the talk is about the way stories can be used to teach technical communicators and to build culture, maybe I should tell stories.

There's the story about how Jerry and I came up with the idea for the book. It had all started about eighteen months earlier . . .

Walking along the shore, I gazed intently at the pockets of gravel in the sandstone shelf to see if I could find an agate, though I had very little idea what to look for. When I found a candidate, I held it up to the sun to see if it looked translucent, supposedly one of the characteristics of agates. Not knowing much about agates, I had already collected half a dozen nicely sea-polished pieces of white quartz. Gazing at yet another piece of quartz in the sun, I heard a voice.

"Dale?"

I looked around, rubbed my eyes, and then decided it was possible, though unlikely—perhaps I had been looking at the sun too long.

"Jerry? What are you doing up here?" Jerry Savage had worked on his Ph.D. at Michigan Tech in the early nineties. Back then, I was an assistant professor, out on my first job after finishing my dissertation. We had become friends, and later, when we both ended up in Illinois, we had continued to keep in touch. Now, here we both were, 500 miles north of Illinois on a little patch of public-access beach in Upper Michigan.

"So it is you, Dale," he responded. "I was watching this guy holding rocks up to the sun, and with the beard and all, I commented to Sue that that guy fits my idea of what Elijah looked like. I was also thinking that he looked curiously like Dale Sullivan, but then you were supposed to be in Illinois for the summer. So how is it that you're on the shores of Lake Superior?"

I explained that my son, Phil, and I had been living in the area for a while during my first year back at Tech, that I was teaching a summer course on C. S. Lewis, and that we were out with friends for a picnic. I pointed to five people in bathing suits lying on the beach as though we were in southern California rather than Upper Michigan. "That's my bunch, the Kuipers and the Anibles. I think you met the Kuipers when we lived up here before. What about you?"

"We're up visiting family, Sue's sister in Lake Linden." Small talk about family and work followed, Jerry talking about his work at Illinois State where he teaches technical communication and takes care of the internship program. He had one more year before he had to go up for tenure, but his publishing was coming along fine. People told him he looked like a good prospect.

I described my first year back at Tech after being gone for six years: the difficulty of living apart from most of the family for a year while we waited for the house to sell; the weight of having given up tenure to make the move and the pressure I was feeling to produce several more refereed articles in the next couple years; the challenge of figuring out how to teach technical communication to students who actually major in tech comm rather than to engineers who simply use writing on the job; my attempts to come up to speed so that I could assume the directorship of the undergraduate program in technical communication in my third year as specified in my contract.

"I taught HU200 this year, twice," I said. "It's a new course, one designed to introduce our students to our degree program and to the profession of technical communication. It's a tough course to figure out."

"What did you try?"

"Well, it didn't seem right to just come up with another general technical writing course since they have to take junior- and senior-level tech writing classes, so we read articles, some from Dan Jones's *Defining Technical Writing,* some from the STC anthology on ethics and technical communication, Sam Dragga's article on ethics from *Technical Communication Quarterly,* a couple of history articles—one by Connors, one by Tebeaux, one by Kynell."

"How did that work?" Jerry, in his quiet, understated way, seemed to really want to know.

"I thought it was great . . . but the students didn't like it. I think the articles are too academic for first-year students and sophomores. I wish I had years of experience as a professional technical writer so that I could play the role of someone teaching apprentices more directly and so that I would have stories to tell about life on the job—you know, the kind of role Perelman and Olbrechts-Tyteca must have had in mind when they said teaching should be seen as a kind of epideictic rhetoric, in which the teacher represents the culture's values and practices to the students. Carolyn Miller said something about tech writing enculturating students, didn't she?"

"That's right, but she was talking about the service course, wasn't she?"

"Yeah, but teaching majors in tech comm really gives me the opportunity to take that notion seriously. I'm closer to the world of professional technical communicators than to the world of engineers and scientists. But still, my students really can't quite imagine what a tech writer does, and I don't know a whole lot more than they do about corporate culture. It's hard to study to be a professional technical communicator when you can't even imagine what one does or how one spends her or his time."

Jerry responded. "It is a different kind of course when you teach tech comm to majors than when you teach it as a service course to engineers, all right. You know," he shifted his weight from one foot to the other and looked down at the beach, "the interns in my program write reports at the end of their internships. Some of those give a pretty good view of what it's like to be a tech writer." He looked up and straight into my eyes. "Maybe I could send you a couple for class."

"That's a good idea," I acknowledged, but I hated the idea of putting Jerry out, sending me stuff that I should have access to myself. Besides, I knew how it would work. He'd do it, and the stuff would come, and I'd put it on the stack with good intentions, and then, after a while I'd forget about it. In ten years I would clean my office and throw the reports away. Not wanting to inconvenience him for something I probably wouldn't use, I added, "But I probably would just lose them. You know, what we really need is some fiction or narrative that features tech writers. Can you think of any tech writers in sitcoms, other than Bob Newhart in his New Hampshire dream? What about books? For science, we have *The Double Helix, The Nemesis Affair, Too Hot to Handle,* and so on. There's Tracy Kidder's *Soul of a New Machine,* but that doesn't talk about tech writers, just about engineers designing a computer."

"How about *Zen and the Art of Motorcycle Maintenance*?" Jerry suggested. "Pirsig was a tech writer."

"Maybe I should read it again, but my main memory is of Pirsig riding across the country on a motorcycle, with his poor son stuck behind him, while he thinks about classical and romantic mindsets."

"How about bringing tech writers in to talk to the class or using the STC chapter?" Jerry was thinking through it, but then he remembered where we were. "Guess Houghton isn't a high-tech center full of tech writers, is it?"

"One way I tried to get a feel for the culture of tech writers was to lurk on the techwhirlers' LISTSERV," I volunteered. "I even made students sign up and lurk too, but it blew them away—way too busy of a list and too focused on technical questions."

"I know what you mean, though," Jerry continued. "There are a lot of good books out there in tech comm, but none that really introduces people to the field in an entertaining way. Guess we'll have to do something about it—seems to be a real need."

An hour later, our conversation over, and Jerry and Sue heading to Sue's sister's place, I drove down the Lac La Belle road past Hermit's Cove, wondering what the odds were that such a meeting could happen and meditating on the role of narrative in teaching technical communication.

A week or so later, Jerry sent me an e-mail expressing his interest in the problem we had been discussing. We began to speculate about how we might do something to meet the need we thought we saw. "How about we put out a call to tech writers? You know, one that starts out something like 'So, ya wanna be a paperback writer, ey?' and then we tell them that we're collecting narratives from tech writers about experiences on the job?" Jerry and I both warmed up to the idea . . .

Yeah, I thought, as I turned off Highway 8 onto 32 in Laona, I might use that story. But why start there? After all, I have a lot of stories to tell about my twenty-year attempt to find out what technical writing is and about what technical writers do. Maybe I could set a theme, something like "the problem of becoming socialized in the field of technical communication," or "the great mystery of technical writing," and then I could plop down the book and its table of contents as the answer to all our problems. I know that's a little overstated, but I really could tell stories about how I have had a need to find out about the field and how I've gradually come to know more and how this book has been really a great help in letting me see the lives and social settings of tech writers.

The sign for Highway 32 has little red arrows on either side, which means that the road travels through a Native American reservation, probably the Oneida Reservation north of Green Bay. I much preferred this road to 141, which runs parallel to it about twenty miles to the east. This road is quiet and goes through some hills and forest, much more relaxing than 141, an overcrowded two-lane road that most people take from Upper Michigan to Green Bay.

What about the story of getting my first job of teaching technical writing? . . .

Having finished my master's in English back in 1979, I couldn't find a job until the beginning of the second semester in the 1979–1980 academic year. Turns out, it was a job teaching technical writing at a two-year college, and I jumped at the opportunity, even though I had never heard of technical writing, never seen a technical writing textbook, never met a technical writer, and didn't have a clue about what engineers and technicians really do. I stayed ahead of my students by two weeks and bluffed my way through. They taught me about the world of engineering through the reports they wrote, and, having inherited a bookshelf full of tech writing textbooks, I read everything I could find.

But I still didn't know anything, so I visited places where, I now understood, technical writers could be found—Boeing Military Airplane Company, NCR, a local civil engineering firm that built water treatment plants in the Mideast. Visiting these places, interviewing managers, reading their documents: these were first steps in discovering the real world of technical communicators. But I was teaching the "service course" in tech writing, a course designed to teach engineers and technicians how to write as part of their engineering work. I wasn't teaching professional technical writers, so I focused more on the textbooks and joined academic societies that had been developed to support teachers of the traditional service course. Most of their members were, like me, frustrated English lit majors with no experience as professional technical writers.

Okay, I thought, that story captures the typical lack of understanding that people often have about the field. Besides, it puts me in the position of the novice and my audience in the position of the experts. No point in trying to strut my professional authority when they know a lot more about the field than I do, and, furthermore, the story fits with the "mystery of technical communication theme." Count that one in. Mental check mark. How much time will I have for stories? After all, I need to talk about the stories we've received and about the value I see them having for teachers and for professionals. My mind began to drift back to the early days of the project . . .

After Jerry and I talked about the book idea that summer on the beach and then by e-mail, we didn't actually do anything. Not then. In October, we met again, this time in Delaware at the Council for Programs in Technical and Scientific Communication. Rehearsing our ideas, we began to get excited about the prospect. Maybe we really could edit a book with stories written by technical communicators. We talked to others at the conference, and receiving encouraging responses, we decided to go ahead with the project. I agreed to send out the call and to collect the drafts of stories as e-mail attachments (.rtf files, please, and no compression). Jerry agreed to start looking for possible publishers.

Late in October, I sent a call out on the techwhirlers' list, asking for stories written by technical communicators, and before long requests for clarification began coming in, and then the stories, mostly from people working in the computer industry. Many of the stories focused on troublesome interactions with subject matter experts (SMEs, a new acronym for me). One in particular, titled "Samurai Review," though written with all the skill of a pro, presented such a devastating view of the maltreatment of technical writers that I worried about its effects on students. Another, by a woman named Weiss, was brilliant literature, but I couldn't figure out how to use it in class, and I wasn't sure if students could relate to the narrator.

I began to have misgivings about the project: Would there be enough of a response to create a collection? Would all of the stories be about writing in the computer industry? Would the emphasis on the struggles that technical communicators go through to achieve recognition make the collection too depressing? . . .

I pulled myself out of my reverie and insisted that it was time to get back to planning the talk. That whole line of thought about troublesome stories brought up one of the functions I hoped the book would perform—giving students looking for a career a chance to consider tech comm. Maybe I could work something into my talk about how hard it is to recruit high school students directly into our program because most of them have never heard of the field. Maybe I could even talk about my own daughter's reluctance to go into the field.

Ember has loved to write ever since first grade and is, in my opinion, gifted at it. But how are you going to make a living as a writer? She knew that writing stories and novels and making a living at it is a long shot, but now she was thinking journalism. Did she have any idea how much (or little) journalists typically make? How much technical writers make? I knew she could earn twice or even three times as much as a technical communicator if she could warm up to the idea, but all she knew about it was that her dad taught it, and it seemed to have a lot to do with computers.

She wasn't interested. Besides, it would be a lot more fun to go away to school. Financial pressure was the determining factor—a tuition break at Tech for faculty family members and cheap living landed Ember in the STC program, but only as a temporary solution until she could figure out how to fund her real dreams.

In January, about three months after the call for stories had gone out, my wife, Sheryl, Ember, and I drove to Marquette to watch our youngest daughter, Becky, compete in a swim meet. I had the manuscripts in my pack, thinking I would have time to read some of the new submissions. Swim meets, I discovered, being a new parent to the sport, are long days of waiting around and a couple minutes of great excitement when your kid gets to swim. Not having yet developed team spirit, I spent most of the day reading and talking to Sheryl and Ember.

The topic turned, as it often did, to Ember's dreams of going away to school and majoring in journalism or elementary ed. It occurred to me that perhaps some of the stories in my pack would change her mind—a couple that I had read that morning were inspiring. So I pulled the file out and suggested that she read a couple while we killed time.

She opened the file, thumbed through, came to "Samurai Review," and asked, "How about this one?" I had read it a couple weeks earlier, and my head was full of the stories I had read that morning. All I could remember on the spur of the moment was that the writer was a really good writer. This one might show her that technical writers can write good stuff, I thought.

"Sure, it's a good story," I said, and then I sat mum on the bleachers, waiting patiently while she read.

"Why would anyone want to be a technical writer?" she asked, turning the last page.

"What do you mean?" I asked, shocked by her unexpected response.

"The way the people at the review session treated that guy was just terrible. If that's what technical communicators have to go through, I don't want anything to do with it."

Too late, I realized that she had read the story to get information rather than to appreciate the writer's skill . . .

Well, I reflected as I slowed to pass through Suring, that really happened, but what does it tell me about the book and the theme of my talk? I guess it brings up questions about why we decided to keep some of the negative stories in the book. As I scanned the countryside, I noted that they didn't have as much snow down here as up north. Now I was in farm country, probably dairy country. More open fields and farmsteads. The forests had been beaten back and tamed around here. But the process of putting the book together was still pretty wild a year ago . . .

Late in February, I e-mailed Jerry, saying I had doubts and asking if he thought it would be worthwhile to continue with the project. He agreed that our collection so far was unbalanced and too small, but he suggested that we stick with it at least until we had a chance to talk about it at the Association of Teachers of Technical Writing conference. We also decided to try to throw our net a little wider. We sent out calls on a couple of other lists and in alumni newsletters. Friends at other universities began to publicize the call as well. Evidently the call found its way to a LISTSERV in England, and someone sent back a cynical note, suggesting that we were playing the part of Tom Sawyer, trying to get other people to paint our fence for us. You never know how people will take things.

A couple months later, we met in Atlanta at a one-day conference of the Association of Teachers of Technical Writing. Jerry, Teresa Kynell, and I were on the same panel, and we were all giving papers on the social aspects of technical communication as a profession.

After giving our papers, Teresa, Jerry, Jerry's wife, Sue, and I had retired to a table in the book exhibitors' area. Teresa Kynell had been planning to put together a book that collected articles about authority and legitimacy in technical communication. Now she was inviting Jerry and me to be coeditors of the collection. As we talked about possible publishers and contributors, I worried inwardly. Jerry and I already had a book project in the works, and I wasn't sure we could pull that off, let alone another book. I told myself to be realistic, to avoid overcommitment, but the atmosphere at conferences has a way of convincing you that all things are possible, and I ended up agreeing to call a few people to ask for chapter contributions. Teresa had phone calls to make, so she excused herself, leaving Jerry, Sue, and me alone at the table.

"I'm really feeling overwhelmed with commitments, Jerry. I don't know if we're going to be able to get our collection of stories together," I said, turning our discussion to our book project. This was hardly an upbeat opening comment, but it wasn't unexpected, considering that I had said the same thing by e-mail a couple of months earlier.

Sue entered the conversation. "I've been reading the stories we have so far," she said, "and I think there is real potential here. You two have come a lot further on this project than you may think. I think all you need is four to six more stories with a little more diversity."

"I've been working on a draft of the proposal to send out," Jerry added. "Allyn and Bacon got in touch with us right after we went public with our intentions, and I think we ought to send it to them."

"What all do they want in a proposal?" I asked. Jerry described the proposal as it now stood and talked about his contact with Sam Dragga, the editor of Allyn and Bacon's series on technical communication.

"I've also been working closely with a couple of former students, encouraging them to send in stories for the collection," he added, "and Teresa says she has contacted a couple of her students, suggesting that they contribute something."

"I can put a call in our alumni newsletter," I admitted, though I doubted that we would get much of a response.

That meeting was enough. I decided to hang on and hope that we could ride this one out. My primary focus had to be on refereed articles in journals, as did Jerry's. We both had tenure to earn, and although we thought this book would be a great contribution to the field of technical communication, we doubted that it would have the status of a refereed, scholarly book. It felt like a labor of love that neither of us had time for . . .

I was now at the intersection where 32 joins 29, a four-lane freeway, the first one I had seen today. Fifteen minutes later, I was circling Green Bay on the freeway, munching on a sandwich from a fast-food place. Soon the traffic died down, and I was able to return to planning the talk as I headed south toward Sheboygan on Interstate 43, cutting across the base of the Door Peninsula. Instead, my mind drifted back to the last year's developments . . .

Last summer I decided I was overcommitted, so I e-mailed Jerry and Teresa and declared my intention to withdraw from being coeditor of the book on authority and legitimacy. Jerry and Sue planned to travel to the Keweenaw Peninsula, where Michigan Tech is located, for a summer vacation in July, so we planned to meet, look at the stories, and decide how to proceed. However, just a week before their planned visit, a family illness kept them from making the trip, and so we decided to continue work by e-mail.

We each pulled up the files that had been sent to us as e-mail attachments, scanned them, and tried to figure out if they constituted a book. There were over twenty stories. Although I had been corresponding with the contributors, I hadn't paid much attention to where they were from. Now I was surprised and elated to see that we had contributions from Canada, from several regions of the United States, from Asia, from Israel. All had been sent as e-mail attachments, and we had already requested revisions from several of the authors. After deciding that there were enough good stories for a small book, we looked for a pattern. Jerry believed he saw the stories falling into three groups, one that consisted of stories focusing on early experiences in the field, one that consisted of stories showing the technical writing process, and one that consisted of stories depicting life beyond the job.

Deciding it was time to send in the proposal, Jerry wrote a final draft and sent it to Sam Dragga. He had already discussed the project with Sam, who lived in Texas, by e-mail. As series editor, Sam read the proposal and then recommended the book to Eben Ludlow at Allyn and Bacon in New York. As we waited, we began editing the stories, deciding to use the comment function in our word-processing software and to track our editing changes so that authors would be able to see both Jerry's and my suggestions. Jerry worked on the table of contents, the preface, and the introductions to the sections. He said it was easy to do because he had most of the material from working up the proposal.

Within a couple weeks of Jerry's sending the proposal, we had an agreement with Allyn and Bacon. I was shocked that things could move so fast. This was not the typical three-to-six-month review of articles I was used to. Now we had to work out the details of the contract and the schedule. Since our contact at Allyn and Bacon was in New York, the most efficient way of discussing these things was by e-mail. Could we have a final draft by December 15? Seemed reasonable to me—that was nearly six months away—but as I drove down I-43, I felt a pang of guilt because we had missed the deadline. Leaving most of the dealings with the publisher to Jerry and Sue, who turned out to be our best in-house legal advisor, I sent the "good news" message out to our contributors, explaining that we had a publisher and asking for snail mail addresses, revisions, photos, and biographies. A flurry of e-mail correspondence followed during the next couple of months, but the fall quarter came upon me with full force—a couple of classes, a couple of conference papers to write, a couple of articles to revise and resubmit, a book review, revision of the STC curriculum—and the book project got squeezed out once again.

A couple of weeks before the end of the fall term, I e-mailed Jerry, assuring him that once my grades were turned in, I would be able to turn my full attention to the book and begin to carry my fair share of the burden. What was left to do? The stories all needed to be edited closely, but reviewers were in the process of reading and commenting on most of the stories, so it would be premature to edit them without the reviewers' comments. The biographies and photographs were beginning to pile up on my disk drive and on my file cabinet, and they would have to be worked in, but the real pressing job was to write my own chapter. The proposal had promised that I would write one, and that chapter was to sort of balance out my textual contribution to the collection, whereas Jerry's contribution was to be the introductions to the sections and the preface to the book.

I decided to be the scholar. For a couple of weeks, sleuthlike, I collected and read articles on narrative. My dissertation had focused on narrative, but that had been ten years earlier. Now I was particularly interested in the literature that was accruing around the topic of how narratives work in organizations to build organizational culture and identity. As I read, I began to see many connections with the stories in our collection.

Mumby explained how stories told in organizations can reinforce or subvert existing hierarchical structures. I thought of stories in our collection, like Lee's "A Job like a Tattoo," in which the author started a new job only to come into conflict with existing prejudices and pecking orders, or like Jong's "Samurai Review," in which the programmers demeaned the technical writer, or of Rasmussen's "Madame Mao in the Midwest," in which the writers found themselves denied the autonomy and respect that professionals expect.

Clair described "sequestered stories," stories about sexual harassment that people usually do not tell publicly. In our collection, Potts tells a story about "Jack," whose comments caused her to worry about his intentions.

Faber wrote about the way stories work to produce change. Certainly, there were stories in the collection that could produce change: Connatser's description of a troublesome internship experience presents a great lesson for both those taking the internship and those directing it; Hile's description of how she helped representatives from a company write their own safety manual seemed to me to have a lot to suggest about how to reconceive the technical communicator's job.

I really liked the comment by Maines and Bridges in another article: they said that stories invite readers to join with the writer and to experience the writer's culture. Who could read the stories by Weiss or Pellar-Kosbar in our collection, for example, and not feel as if they had really met these people and entered their lives and culture?

Blyler's article discussed the role stories play in initiating members into organizations, and, similarly, the one by Ledwell-Brown and Dias claimed that stories are useful in teaching how to operate in a new culture. I realized that the whole collection of stories could be seen as a way of bringing students into the culture of the profession. I was anxious to use the collection in my classes.

Now, how much of that theory stuff should I bring into the talk tonight? I asked myself as I passed Sheboygan. I don't have a lot of the details with me. Better gloss over it and just make general broad comments. There is that one slide with quotes from theorists on narrative. I'll use that after we go through the table of contents.

There is, of course, the angle that these stories could serve as substitute ethnographic studies. I thought about my doctoral studies at Rensselaer Polytechnic Institute. Back then, people were beginning to take seriously the mandate to find out how writing works in the real world. Lee Odell and Dixie Goswami had published a collection of ethnographic studies of writing in nonacademic settings, the title itself showing just how much we still thought of academia as the center of the world, with everything else being defined by its status as "nonacademic." Carolyn Matalene, and later, Rachel Spilka had published collections of ethnographic studies of workplace writing. Several doctoral dissertations had been ethnographic studies of writing in industry, hospitals, or government, but I couldn't think of any that were ethnographic studies of professional technical communicators. Maybe I could claim that these studies performed some of the same functions as ethnographic research . . .

■ ■ ■

That had been Tuesday, headed south. Today, I was headed north, passing Sheboygan, Green Bay in my sights. I smiled to myself. It's remarkable how these things turn out, I thought . . .

We had met in a room that could seat about sixty people. We ate finger food and talked informally, sitting at round tables big enough for six or seven people. It was obvious that many of the people knew each other: they updated one another on projects, asked about family members, described new positions that they had taken. Some were there for the first time, and I could see that they were beginning to realize what a wealth of information was collected in one room. They tracked down possible job openings and professional contacts, discussed new software, and exchanged business cards. After an hour, Mollye moved to the front of the room and introduced me.

With a little nervousness about how these real technical communicators would take to an academic telling them about their profession, I opened with praise for Milwaukee and suggested that we in Houghton liked to visit the suburbs (Minneapolis and Milwaukee) now and then, just to see how things were going. Then I moved into the talk by telling my stories and summarizing some of the plots of the submitted stories.

When I told them the story about Ember reading "Samurai Review," a vigorous discussion broke out about how the sharing of "horror stories" has become standard practice when tech writers get together. I assured them the story of the review was fiction.

"No, that's real life," someone quipped. Laughter and nodding heads.

"Why is the sharing of such stories useful?" I asked. Several theories came in response: they serve as negative examples to help us learn how to avoid situations like them; they build a sense of professionalism because we are motivated to put an end to people ignoring our expertise; they make us feel part of the culture because we all have some horror story to tell.

"It would be inaccurate to put out a book representing technical communicators if all the stories were positive," someone observed. "Students need to know that there is still headway to be made."

I came to Part III of the book. "These stories are about people with lives outside the job," I began.

"Now that's fiction!" I heard from my right about halfway back. More laughter. They were warming up, enjoying the summaries of the stories, obviously identifying with some of them.

I mentioned that some of our reviewers had suggested that we get rid of the negative stories and references to specific technology. The negative stories might be too hard for students to swallow, and the technology would date our book. There were so many people wanting to get in on these topics that the discussion lasted a little past quitting time.

"No, no. Leave both in. The horror stories are part of our culture."

"How about a volume with stories and then commentary by practitioners, suggesting how the situation might have been handled more effectively?"

"Good idea," I responded, "I hadn't thought of that." I marveled, realizing that there were lots of ways to work with narratives in the field.

"Leave the references to technology in. The book will serve as a kind of archive for the evolution of technical communication practice."

"Yeah, you need to produce more than one book, so that we have a record."

"I have stories I could submit."

"Keep giving voice to the profession."

Mollye closed the meeting with a comment that made me feel that it had gone well: "We've been mentored tonight by someone who really isn't a technical writer. That doesn't happen very often, and it's a pleasure to have had the experience."

It had gone better than I expected. The meeting had paid off for me in ways I hadn't foreseen. I met technical writers who worked for Compuware; the new manager for a technical writing contracting company named Interim, who had, incidentally, just published a children's book about a girl with a fountain in her head; Ken Cook, a former president of STC; and the manager for Knowledge Management Services at Iverson, Mollye, who had run the meeting and made the arrangements. All of these people were potentially valuable contacts for our students. STC turned out, once again, to be a great place to "network."

Since it was Mollye who had contacted me to see if I would be willing to speak to the chapter, she drove me to and from the meeting. On the way back to County Clare Inn, we talked about socializing new tech writers. I mentioned our need to create a better network for our students to participate in co-ops and internships. Mollye said that she sometimes sends new tech writers out to shadow veterans on the job for half a day. She thought it would be a good experience for some of our majors in technical communication at Michigan Tech.

"How many students do you think members of the chapter would host, if I brought some down?" I had asked.

"Oh, it wouldn't be hard to find at least ten people who would be happy to show a student around for half a day."

"Wow! That would be great. We oughta do that. Let me think about it and get back in touch. But maybe we don't have to do it right away."

"Always do it now, if you're going to do it," she responded . . .

The engine hummed away. Only five hours to Houghton. Maybe I can make it back in time to hear about the Irish immigrants to the Copper Country. So, what's left to do? One more story to edit. Time to throw away that chapter I wrote: I know what I can do now. Better get an e-mail out to the majors to see if any are interested in a field trip to Milwaukee. The proposals for the new certificates in our program need to be sent to the steering committee. The engineering enterprise project has waited long enough; I promised Patty that it was next on my list of top priority projects . . .

Further Reading on Narrative and Writing in the Workplace

Barton, Ben F., and Marthalee S. Barton. "Narration in Technical Communication." *Journal of Business and Technical Communication* 2 (1988): 36–48.
Bennett, Lance W., and Murray Edelman. "Toward a New Political Narrative." *Journal of Communication* 36 (1985): 156–171.
Blyler, Nancy R. "Narration and Knowledge in Direct Solicitations." *Technical Communication Quarterly* 1 (1992): 59–72.
———. "Pedagogy and Social Action: A Role for Narrative in Professional Communication." *Journal of Business and Technical Communication* 9 (1995): 289–320.
Boje, D. M. "The Storytelling Organization: A Study of Story Performance in an Office-Supply Firm." *Administrative Science Quarterly* 36 (1991): 106–126.
Bormann, Ernest G. "Symbolic Convergence Theory: A Communication Formulation." *Journal of Communication* 36 (1985): 128–138.
Brodkey, Linda. "Writing Ethnographic Narratives." *Written Communication* 4 (1987): 25–50.
Bruner, Jerome. "The Narrative Construction of Reality." *Critical Inquiry* 18 (1991): 1–21.
Clair, Robin P. "The Use of Framing Devices to Sequester Organizational Narratives: Hegemony and Harassment." *Communication Monographs* 60 (1993): 113–136.
Conrad, C. "Organizational Power: Faces and Symbolic Forms." Ed. L. Putnam and M. Pacanowsky. *Communication and Organizations: An Interpretive Approach.* Beverly Hills, CA: Sage Publications, Inc., 1983. 173–194.
Faber, Brenton. "Toward a Rhetoric of Change: Reconstructing Image and Narrative in Distressed Organizations." *Journal of Business and Technical Communication* 12 (1998): 217–237.
Fisher, Walter R. *Human Communication as Narration: Toward a Philosophy of Reason, Value, and Action.* Columbia: U of South Carolina P, 1987.

Kermode, Frank. "Secrets and Narrative Sequence." Ed. W. J. T. Mitchell. *On Narrative.* Chicago: U of Chicago P, 1980. 79–97.

Ledwell-Brown, Jane, and Patrick X. Dias. "The Way We Do Things Here: The Significance of Narratives in Research Interviews." *Journal of Business and Technical Communication* 8 (1994): 165–185.

Lucaites, John L., and Celeste M. Condit. "Reconstructing Narrative Theory: A Functional Perspective." *Journal of Communication* 36 (1985): 90–108.

MacIntyre, Alasdair. *After Virtue: A Study in Moral Theory.* Second edition. Notre Dame: U of Notre Dame P, 1984.

Maines, David R., and Jeffrey C. Bridger. "Narratives, Community, and Land Use Decisions." *The Social Science Journal* 29 (1992): 363–380.

Martin, Joanne. "Stories and Scripts in Organizational Settings." *Cognitive Social Psychology.* Ed. A. Hastdorf and A. Isen. New York: Elsevier/North-Holland, 1982. 255–305.

Martin, J., M. Feldman, M. J. Hatch, and S. B. Sitkin. "The Uniqueness Paradox in Organizational Stories." *Administrative Science Quarterly* 28 (1983): 438–453.

Matalene, Carolyn B. *Worlds of Writing: Teaching and Learning in Discourse Communities of Work.* New York: Random House, 1989.

Mitroff, Ian I., and Ralph H. Kilmann. "Stories Managers Tell: A New Tool for Organizational Problem Solving." *Management Review* 64 (1975): 18–28.

Mumby, Dennis K. "The Political Function of Narrative in Organizations." *Communication Monographs* 54 (1987): 113–127.

Odell, Lee, and Dixie Goswami. *Writing in Nonacademic Settings.* New York: Guilford Press, 1985.

Rosen, M. "Breakfast at Spiro's: Dramaturgy and Dominance." *Journal of Management* 11 (1985): 31–48.

———. "You Asked for It: Christmas at the Bosses' Expense." *Journal of Management Studies* 25 (1988): 463–480.

Spilka, Rachel. *Writing in the Workplace: New Research Perspectives.* Carbondale: Southern Illinois U P, 1993.

Sullivan, Dale L. "Exclusionary Epideictic: *NOVA's* Narrative Excommunication of Fleischmann and Pons." *Science, Technology, & Human Values* 19 (1994): 283–306.

Turner, Victor. "Social Dramas and Stories about Them." Ed. W. J. T. Mitchell. *On Narrative.* Chicago: U of Chicago P, 1980. 137–164.

White, Hayden. "The Value of Narrativity in the Representation of Reality." Ed. W. J. T. Mitchell. *On Narrative.* Chicago: U of Chicago P, 1980. 1–23.

Witten, Marsha. "Narrative and the Culture of Obedience at the Workplace." Ed. D. K. Mumby. *Narrative and Social Control: Critical Perspectives.* Newbury Park, CA: Sage Publications, Inc., 1993. 97–118.